T0234459

PHILOSOPHICAL FOUNDATIONS FOR THE PRACTICES OF ECOLOGY

Ecologists use a remarkable range of methods and techniques to understand complex, inherently variable, and functionally diverse entities and processes across a staggering range of spatial, temporal, and interactive scales. These multiple perspectives make ecology very different from the exemplar of science so often presented by philosophers.

In this book, designed for graduate students and researchers, ecology is put into a new philosophical framework that engages with this inherent pluralism while still placing constraints on the ways that we can investigate and understand nature. The authors begin by exploring the sources of variety in the practice of ecology and how these have led to the current conceptual confusion. They argue that the solution is to adopt the approach of *constrained perspectivism* and go on to explore the ontological, metaphysical, and epistemological aspects of this position and how it can be used in ecological research and teaching.

BILL REINERS has practiced ecology for 45 years. His practice has primarily been focused at the ecosystem level, more specifically with biogeochemical phenomena, but he has also been interested in vegetation patterns across landscapes and the nature of disturbance and recovery in vegetation and ecosystems. His teaching and research have taken him to many terrestrial habitats ranging from tropical and temperate rain forests, temperate deciduous and coniferous forests, and sagebrush steppe. He is an ISI Highly Cited researcher, has been recognized with a number of campus-wide awards at the University of Wyoming, and as a distinguished alumnus of Rutgers University.

JEFF LOCKWOOD has practiced ecology for 25 years. After studying behavioral ecology (semiochemical communication), he became internationally recognized for his work on grasshopper management, pioneering a method that reduced insecticide use by >50% in the western USA. He was among the first to apply complexity science to model the population dynamics of insects. Fascinated by the interface between the natural sciences and humanities, he now enjoys a joint appointment at the University of Wyoming between the department of philosophy and the MFA program in creative writing. His writing has been recognized with a Pushcart Prize and a John Burroughs Award.

Philosophical Foundations for the Practices of Ecology

WILLIAM A. REINERS

JEFFREY A. LOCKWOOD

CAMBRIDGE
UNIVERSITY PRESS

CAMBRIDGE
UNIVERSITY PRESS

University Printing House, Cambridge CB2 8BS, United Kingdom

One Liberty Plaza, 20th Floor, New York, NY 10006, USA

477 Williamstown Road, Port Melbourne, VIC 3207, Australia

314-321, 3rd Floor, Plot 3, Splendor Forum, Jasola District Centre, New Delhi - 110025, India

79 Anson Road, #06-04/06, Singapore 079906

Cambridge University Press is part of the University of Cambridge.

It furthers the University's mission by disseminating knowledge in the pursuit of education, learning and research at the highest international levels of excellence.

www.cambridge.org
Information on this title: www.cambridge.org/9780521133036

First published 2010

A catalogue record for this publication is available from the British Library

Library of Congress Cataloging in Publication data
Reiners, William A.
Philosophical foundations for the practices of ecology / William Reiners, Jeffrey Lockwood.
p. cm.
Includes bibliographical references and index.
ISBN 978-0-521-11569-8 (Hardback) – ISBN 978-0-521-13303-6 (Pbk.)
1. Ecology–Philosophy. 2. Perspective (Philosophy) I. Lockwood, Jeffrey Alan, 1960– II. Title.
QH540.5.R45 2009
577.01–dc22
2009024548

ISBN 978-0-521-11569-8 Hardback
ISBN 978-0-521-13303-6 Paperback

Contents

Preface

Our lives are chronologically linear, but are experientially complex. As Søren Kierkegaard said, "Life must be lived forward, but can only be understood backwards." What we are and how we think are products of countless collisions with the influences of others: parents, teachers, writers, artists and colleagues. Of course, we are not like mere gas molecules entrapped in a vessel, colliding randomly with one another. Rather, our histories produce an accumulated trajectory – an intentional trajectory that is knocked about, diverted, altered in velocity and direction by encounters with new influences. As with ecological systems, initial conditions are critical, and geography and culture are constraints, but history still matters to our life courses. This book results from the interaction of two ecologists who have different starting times and geographies, as well as career trajectories pointing in different directions.

I (Reiners) was born at the end of the depression, remember WWII, comfortably surfed the front of the baby-boomer wave throughout life, was nurtured in a small liberal arts college, and followed an ecological career through plant ecology, ecosystem ecology, biogeochemistry, and spatial applications. In retrospect, my career seems to have been an evolutionary path along which ecological perspectives were continually undergoing modification. The fact that I moved around, from Illinois to New Jersey to Minnesota to New Hampshire to Wyoming, and that I worked in Alaska and Costa Rica was probably important to my continual reframing of ecological concepts in different ways. At the same time, my attitudes toward objectivity were naïve, my thoughts on motivation unfocused, and understanding of philosophy weak – possibly conventional traits of my generation. There is no way I could have contributed to a book like this without having encountered Jeff Lockwood – an extraordinarily collision of trajectories.

I (Lockwood) am a baby boomer, remember the Vietnam War, attended a small, technical institution followed by a big state university, and traced an academic career that took me from animal physiology, to insect behavior, to grasshopper ecology and management, and finally to the arts and humanities. My professional life has been spent at the University of Wyoming, but I've moved around a great deal in intellectual terms. I began with behavioral ecology, explored complexity models of insect population dynamics, and developed new methods – with dramatically fewer economic and environmental costs – for killing grasshoppers. But I always felt a pull toward the interface of the natural sciences and humanities and finally metamorphosed into my current split appointment between the department of philosophy and the MFA program in creative writing. My peripatetic scholarship is such that I lacked the ecological depth to pull off a book of this sort, which made my pleasant collision with Bill Reiners essential to what you are now holding.

How we came to write this book is described in Chapter 1. We were both having trouble rationalizing ecology as a discipline from a philosophical point of view and turned to a graduate seminar as a means to thrash things out. That seminar opened up the extent and nature of philosophical problems that were troubling us, but it took us several years to formulate the positions we present in this book.

The sequential organization of this book is represented by the chapter titles. We explain how we came to address these issues in Chapter 1. We describe our perception of the philosophical confusion that pervades ecology in Chapter 2 and explore the causes in Chapter 3. We begin to put ecology into a philosophical framework with an imaginary "walk in the park" in Chapter 4. This philosophical framework is expanded and filled-in, and a philosophy of constrained perspectivism developed in Chapters 5 through 9. Chapter 10 presents our interpretation of philosophical positions found in the practices of ecology historically and into the present, with examples from our own personal experiences. We suggest that for the most part, we ecologists don't have much of an idea of what we are about with respect to philosophy and that this has unfortunate consequences for us.

What to do about this? We individually can pull ourselves up by our bootstraps, but it is more critical to consider the philosophical implications of how and what we teach in ecology. It is our successor generations with which we should be most concerned. We address the implicit philosophies underpinning our teaching of undergraduates and graduate students in Chapter 11, making suggestions for how this might be improved, and what implications improvement might have for those of us well along in our careers. Finally, in Chapter 12, we return to how we ecologists function as scientists with or without much concern for guiding philosophies, and how we are not, in fact, so dysfunctional for all of our confusion. We hope that readers will at least be amused, if not accepting, of our mythic image of the ecologist as not Sir Isaac Newton, but as Joe the Plumber – a resourceful handyman.

Inasmuch as we espouse a pragmatic and pluralistic constrained perspectivism, we expect and accept that readers may find this book disturbing and unacceptable. We see that as our fate, but nevertheless hope that this may be a contribution toward the maturation of our discipline, if not in this generation, in the next.

Acknowledgements

To all of our colleagues at the University of Wyoming with whom we have discussed science in general, and ecology in particular; and who provided advice on our title and their reactions to the photo in Box 2.1, we offer our sincere appreciation. While our colleagues surely will not align themselves with all that we say, we still travel in a common quest for understanding and wisdom as a community of scholars. We are fortunate to be members of this community.

Among our University of Wyoming colleagues, special thanks go to Franz-Peter Griesmeier, Department of Philosophy, for a critical reading of an earlier version of this manuscript, and for his expert consultation on science philosophy. Franz-Peter may not have accepted all of what we say, but he understood and constructively critiqued our perspective – all that we ask of any reader. Contributors from the Department of Botany included Kenneth L. Driese who helped with photo processing, Daniel B. Tinker who provided insight on attitudes toward wildfire, and Naomi L.Ward who showed us current understanding of pangenotypes. Karen Bartsch, Department of Psychology, provided critical information on current thinking about the cognitive development of students.

We benefited from many valuable conversations on the challenge of teaching ecology with several colleagues within and outside of the University of Wyoming, including Robert E. Ricklefs, University of Missouri at St. Louis; Daniel Simberloff, University of Tennessee; Margaret D. Lowman, New College of Florida; Robert O. Hall, University of Wyoming; Rebecca E. Irwin, Dartmouth College; Manuel C. Molles, Jr, University of New Mexico; and Mitchell Pavao-Zuckerman, University of Arizona.

Our warm appreciation is extended to our many students over the years, but especially the outstanding collection of graduate students who responded to our invitation to explore with us the nature of science philosophy as practiced in ecology. We fondly remember the open-minded intensity and intellectual honesty of this particular group of students who provided the fertile soil from which this book grew. Thank you and may your receptivity to many – but not just any – perspectives stay with you throughout long and fulfilling careers in ecology.

This book would not have been possible without the moral and intellectual support of our wives, Norma and Nancy, who understood how important this endeavor was to us and gave us their unflagging encouragement.

Finally, we thank Dominic Lewis of Cambridge University Press for his prompt and helpful efforts to bring this book to fruition.

1

Introduction

The seed for this book was planted in 2005 when we organized an *ad hoc* graduate seminar on ecology's philosophy of science. Earlier experiences had demonstrated how such seminars, rich in readings and discussions, were effective ways to explore unfamiliar and complex topics. After we announced the seminar, some 15 students joined us, and as soon became clear, they were as uneasy and puzzled about philosophical problems and solutions in ecology as we were.

What were some of these questions that drove our explorations at the interface between ecology and philosophy? First, what was the nature of a scientific philosophy associated with ecology – if there was one? Ecology itself seemed to be disintegrating into ever more sub-fields with fragments reintegrating into meta-sciences like earth system science, conservation biology, and so forth. Did anything "hold the center" of ecology? Could there be a single philosophy for all of our science, and if there were many philosophies, were they coherent with one another?

Second, what place did laws or theories hold in ecology? In fact, semantic usage of these terms was confusing. "Theory," "law," "paradigm," and "model" clearly meant different things to different people. Maybe after we clarified the language we would discover what these terms meant to ecologists.

Third, in the apparent absence of widely accepted generalizations about the nature of the world, how could we determine whether ecology was actually progressing in its quest to understand that part of nature it addresses? And what constituted progress? Were ecologists seeking objective truth, simplicity, unification, coherence, human well-being, or some other objective against which improvement could be assessed? To be sure, ecology had accumulated an enormous understanding of the details of particular situations and phenomena: how isolated functions

1

worked, the imprint of history, the importance of spatial relations, and the criticality of scale definition. Aside from this mass of information, though, was ecology actually assembling a set of general theories or rules that could be universally applied – and to what ends? For some of us, most of the more current conceptualizations of nature seemed to be merely flashy jargon supported by *avant garde* mathematical, statistical and analytical methods that, with some historical perspective, simply echoed older views. Had we achieved a more profound (as opposed to detailed) understanding of nature than Forbes described in 1887?[1] If so, what was the basis for this claim?

Fourth, we realized that standard notions and terminology of philosophy were largely missing from ecological debates. This suggested that there really was little explicit connection between science – at least this science – and philosophy. But was such an association important or useful?

As the seminar progressed, we came to realize and grudgingly admit that our scientific activities were driven by our human needs and wants, as well as personal interests. This too was a part of ecological philosophy. In spite of our high-minded rationalizations of what we do, we had to admit we explored nature along paths we preferred for sometimes unknown (or at least unexamined), and often non-rational reasons. Some of these motivations derived from our individual cognitive development, cultural fixations, sense of natural aesthetics, or ethical impulses. Very often, funding availability set our ecological agendas.

We asked ourselves – and invite readers to do the same – these questions:

(1) Why did we choose to become ecologists? Was it the intellectual qualities of the material, the lifestyle, the social approbation of the field, the possible relevance to stewardship of Earth, the beauty and value of organisms and their interactions, the potential to ameliorate human suffering, or something else? What affect did these motivations have on what we chose to study and claimed to believe?

(2) Are the things we study real, or did we just make up some of them? Is nature truly composed of populations, organized into communities, and related to the physical universe as ecosystems? Are these entities really demonstrable? Does it matter if ecology is a useful fiction? Are there limits to what we can create?

(3) If we are serious about the reality of entities and processes we study, then what standards of evidence do we require to test the truth of what we think? On what grounds can we say yes, here is the evidence for a population, community or ecosystem and its properties? Is there

more than one legitimate way to perceive reality? Is just any way allowable or are there constraints?

(4) Are the things that we believe to be true, just true under certain circumstances, or are any of them universally true? If they are only contingently true, what rules do we follow to relate the necessary auxiliary information to apply a "law" to a particular case? And exactly what do we mean by something being true?

(5) Is there just one truth about how nature *really* is – an ecological "holy grail" – or can there be more than one? Are truths relative to the domains in which they exist or can we access absolute Truth? Can some truths be only partial, but still have great value? Is it possible to know the truth or are we so burdened by our own cultural and individual conditioning that we may only get glimpses through distorted personal lenses? Is the discovery of objective truth the best or only purpose of ecology?

We didn't have to look far to find expressions of confusion, criticism, cynicism, angst, and counter-angst in the ecological literature (Box 1.1).[2] We weren't alone in our suspicions. Many of the voices we read were harshly critical and often contradicted one another.

At the conclusion of the seminar, the students returned to their courses, teaching, and degree research; doing what they were paid – and paying – to do. Time will tell whether the seminar experience has lasting meaning for them. We, the putative instructors of the course, remained so intrigued by this experience, however, that we continued to read, discuss with others, and shape our tentative conclusions into essays. Those essays finally led to this book.

Our book is organized into six sections, each representing a specific purpose. These objectives are:

(1) to illustrate the ambiguity, conflict, confusion, and vagueness[3] within the ecological community about the intellectual basis of our pursuit of useful knowledge and "truth";

(2) to explore reasons why ecology is especially vulnerable to such uncertainties;

(3) to introduce the basic framework of philosophy, in general, and the pursuit of truth and knowledge, in particular;

(4) to introduce a conceptual framework that both accounts for current practices and advocates creating a sound basis for what ecologists ought to believe is valuable, known and real, and how they can justify their claims to know what is true about nature;

Box 1.1
Critical statements about ecology

*"ecology is awash with all manner of untested (and often untestable)
models ... many simply elaborations of earlier untested models."*
(Simberloff 1980: 52).

Speaking of fields including ecology: *"goals and criteria are poorly
enunciated, less accepted, and more sporadically applied. These sciences
are less coherent, they contain many constructs of dubious merit,
and their growth is lethargic."* (Peters 1991: 1).

Paraphrasing a listing of short-comings: *"1) lack of rigour, 2) weak
predictive capability, 3) failure to harness modern technology."*
(DiCastri and Hadley 1986: 300).

"ecology is dominated by complex and inadequately undefined [sic]
*terms which confound the development of predictive theory. As a result,
ecological classifications, ecological characteristics and ecological
relationships may refer to phenomena that vary with each change in
focus, scale, or author, and ecologists are often not sure they are talking
about the same thing."* (Peters 1991: 104).

*"[The] attempt to establish ecology as a mechanistic science like physics
... is self-defeating."* (Keller and Golley 2000: 320).

*"much of ecology is confused in its goals, uncertain of its strengths,
and inconsistent in its terminology."* (Rigler and Peters 1995: 77).

*"We shall argue that, insofar as ecology is required for solving practical
environmental problems, it is more a science of case studies and
statistical regularities, than a science of exceptionless, general laws."*
(Schrader-Frechette and McCoy 1993: 1).

*"We are practicing ecologists. We are not statisticians, numerical
analysts, or philosophers."* (Hilborn and Mangel 1997: 11).

*"Enough of ... poking ecological material to see what happens in some
general way. Ecology needs to identify the critical points of tension, and
then the empiricists need to test predictions coming from explicit theory."*
(Allen and Hoekstra 1992: 332).

*"Ecology is replete with dichotomous debates, divergent scales,
causal alternatives, and conceptual difficulties that can be solved by
integration."* (Pickett *et al.* 2007: 24).

(5) to suggest ways that ecologists can come to terms with explanatory and, to the extent possible, predictive theory in a contingency-laden, heterogeneous, changing world; and

(6) to consider the nature of what we teach, the state of cognition and attitude of our students, and how our teaching serves as a kind of intellectual template that shapes ways of knowing into a common form across the generations.

We know that ecologists don't care about philosophy very much. Even in the most academic variants of self-defined basic research we are uncomfortable and impatient with philosophy. Ecologists may well ask (and do): does this philosophizing matter – or is it just pedantic distraction lying between us and doing "good science"? As practical as ecologists tend to be, it is easy for established scientists to continue to work comfortably in their studiously guarded domains, continuing to do what assures professional advancement (i.e. generates contracts, external funding and refereed publications). As such, near-consensus concerning philosophical skepticism is to be expected given the sociology of science. We are all in this together – participating in a culture that reinforces the same, familiar, and comfortable ways of asking questions, securing funds, performing research, and valuing ideas. This may be especially true with government-sponsored, basic research. The norms of science become self-reinforcing as successful researchers become agency panelists, program directors, National Research Council committee members, book editors, and societal officers. It is natural, then, for us to become entrenched via positive feedback loops in this successful mode of doing business. In that context, we can understand the irony that new or unusual phenomena external to the human institution of ecology (e.g. acid rain, exotic species invasions, disease and pest epidemics, epic wildfires, climate change, and requirements for ecological restoration) dramatize the weaknesses of ecology as a science on one hand, and stimulate it to be a better science on the other. Applied ecology is good medicine for basic science. There's nothing like a dose of empiricism to treat a theoretical malaise – and nothing like a jolt of economics and politics to stimulate inquiry.[4]

But we return to the question, "Does philosophizing matter?" Whereas ecologists can, and do, respond negatively, pointing to success as internally defined by the culture of which we are a part (many "productive" scientists are utterly disassociated, intellectually if not operationally, from philosophical matters), we insist that philosophy always matters and we should do better. If we are to be scholars, rather than technicians (which is not to disparage technicians, who are essential to the practice of ecology), then it is incumbent upon us to think deeply and critically about the nature of our work. Surely a university education, let alone a Ph.D., means more than mere methodological competence.[5] And we will insist that a grasp of philosophy is no less (and perhaps more) important if we teach.

Where students – those outside our culture peering in – encounter our worldview(s), we experience a revealing disjunction between our way of seeing nature and seeking evidence to inform ourselves, and theirs. Of course, we see a central task of our teaching as simply informing students of our way of perceiving nature. But, as those who teach general ecology to undergraduates know well, our multiple approaches to understanding nature and our fuzzy systems of theories and laws seem incoherent, or at least disconcerting, to our "naïve" students. These innocents see that the emperor has an elaborately woven fabric of rhetoric, but suspect that such obfuscation is surely evidence of nakedness. Teachers may interpret this as the students' failure to learn a complex and nuanced science, but it could be our failure to appreciate the students' philosophical realization. Just as a child recognized the emperor had no clothes, our students' more nascent, uncommitted states allow them to see through the cloak of our authority (we write the texts and exams) into the seemingly incoherent ideas of what ecologists contend is real and how they claim to know.

We propose in this book a philosophical position that is "descriptive" of our perception of the contemporary, American ecological endeavor. This description is explicitly tolerant and intentionally pluralistic, albeit not without limits. On the other hand, this position is also "prescriptive," in maintaining that we ought to recognize and tolerate different ways – but not just any ways – of understanding nature. This position requires that we allow the possibility of multiple truths and that we seek value in different ways of knowing nature, but it also demands that we judge with rigor and discrimination the evidence used to support claims about truth and knowledge. We term this position "constrained perspectivism."

Contrary to the sense we may have conveyed to this point that ecologists are philosophically ambivalent or even averse, we have come to believe that the position of constrained perspectivism is the most widespread philosophy that American ecologists hold, even if they aren't entirely conscious of it. And, as we shall see, this lack of conceptual clarity has led us into unproductive conflicts which might well have been avoided – or more effectively resolved – if the disputants had been able to see the philosophical roots of their positions. In a sense, we seek to make the implicit explicit, so that ecologists can more effectively and confidently pursue their research and teaching. This task is analogous to moving from a vague idea, to a graphical representation, and then to a mathematical expression. When we become more precise in our understanding of nature – or science – we can appreciate the limits and opportunities of a system. Our philosophical prescription should be non-threatening to the vast majority of ecologists and, indeed, helpful to their understanding of their own beliefs and behaviors – as subconscious as those might be. It seems to us that while ecologists do not share an acknowledged "conventional wisdom" (philosophy) about their discipline, they do share a "conventional intuition." Many will read this and say, "Why of course – this is how we do it. This is not news!" In this regard, William James, the great American pragmatist philosopher, wrote:[6]

> I fully expect to see the pragmatist view of truth run through the classic stages of a theory's career. First, you know, a new theory is attacked as absurd; then it is admitted to be true, but obvious and insignificant; finally it is seen to be so important that its adversaries claim that they themselves discovered it.

Some may find "constrained perspectivism" (which is derived from pragmatism) to be absurd, but we think these will be few. Rather, we expect that many will find it to be obvious once it is explicated, as might be expected of an account that accords with one's intuitions and experiences – even if the precise terms and form of the account weren't quite self-evident.[7] As for adversaries, and we are not so naïve as to believe that our philosophy of ecology will be without its critics, it may be a bit too optimistic to expect that they will embrace our views as being their own – at least within our professional lifetimes. In the end, we hope that this book will clarify for ecologists how we mostly think and behave, and allow us to better pursue our understanding of nature.

Endnotes

1. Forbes, S. A. 1925. The lake as a microcosm. *Illinois Natural History Survey Bulletin* **15**: 537–550. Reprint of original publication in 1887 from the *Bulletin of the Scientific Association (Peoria, IL)* (77–87).
2. Simberloff, D. 1980. The sick science of ecology: symptoms, diagnosis, and prescription. *Eidema* **1**: 49–54; Peters, R. G. 1991. *A Critique for Ecology*. Cambridge, UK: Cambridge University Press, 1; De Castri, F. and M. Hadley. 1986. Enhancing the credibility of ecology: is interdisciplinary research for land use planning useful? *GeoJournal* **13**: 299–325; Peters, *A Critique for Ecology*, 104; Keller, D. R. and F. B. Golley. 2000. *The Philosophy of Ecology: From Science to Synthesis*. Athens, GA: University of Georgia Press, 320; Rigler, F. H.

and R. Peters. 1995. *Science and Limnology*. Oldendorf/Luhe: Ecology
Institute, 77; Schrader-Frechette, K. S. and E. D. McCoy. 1993. *Method in
Ecology: Strategies for Conservation*. Cambridge, UK: Cambridge University
Press, 1; Hilborn, R. and M. Mangel. 1997. *The Ecological Detective:
Confronting Models with Data*. Princeton, NJ: Princeton University Press, 11;
Allen, T. F. H. and T. W. Hoekstra. 1992. *Toward a Unified Ecology*. New
York: Columbia University Press, 332; Pickett, S. T. A., J. Kolasa and
C. G. Jones. 2007. *Ecological Understanding: The Nature of Theory, the Theory
of Nature*, 2nd ed. San Diego, CA: Academic Press, 24.

3. McNeill, D. and P. Freiberger. 1993. *Fuzzy Logic*. New York: Simon and
    Schuster.

4. Wimsatt, W. C. 2007. *Re-engineering Philosophy for Limited Beings*.
    Cambridge, MA: Harvard University Press, 6. "this view of science and nature
    [re-engineering] is constructed largely (as with all creative acts) by taking,
    modifying, and reassessing what is at hand, and employing it in new contexts,
    thus *re*-engineering. . . . And any engineering project must be responsive to real
    world constraints, thus realism. Our social, cognitive, and cultural ways of
    being are no less real than the rest of the natural world, and all together leave
    their marks. But putting our feet firmly in the natural world is not enough.
    Natural scientists have long privileged the "more fundamental" ends of
    their scientific hierarchy, and pure science over applied – supposing that
    (in principle) all knowledge flowed from their end of the investigative
    enterprise. Not so."

5. Ibid, 26. "An adequate philosophy of science should have normative force.
    It should help us to do science or, more likely, to find and help us avoid sources
    of error, since scientific methodologies are by nature open-ended."

6. James, W. 2006 (1906). What pragmatism means. In *Pragmatism Old & New:
    Selected Writings*, ed. S. Haack. Amherst: Prometheus, 289–308.

7. Wimsatt, *Re-engineering Philosophy*, 8. Constrained perspectivism may be a
    more formal description of Wimsatt's "heuristic techniques" involving scarcely
    conscious mental judgments we all make to understand how things work.

2

Conceptual confusion in ecology

Ecologists understand the nature of science, of ecology, and of nature itself in various ways. Why is there such variety? And does it matter? Should this heterogeneity be seen as a regrettable but necessary source of confusion, symptomatic of an underdeveloped discipline? Should it cause a sense of intellectual inadequacy or even anguish and stimulate efforts to unify or homogenize our understandings? Or should it be recognized as a necessary and natural source of strength in the science? In this chapter, we explore the sources of variety at a philosophical level and how variety in both perception and practice creates confusion and controversy on one hand, but also represents the raw material of ecology and the efficacious products of our science on the other. We also explore whether these questions might better be considered as having value to ecologists when understood in terms of the purposes that motivate them, rather than being rejected because of the ways that they differ from our normal, entrenched, or monolithic perspectives.

The roots of confusion

Even if we are individually certain about the philosophical underpinnings of what we do, it should be evident from a walk down the hall of a biology or ecology department in the average American university that there is little consensus in how the natural world is seen or how the science of ecology is practiced. If one were to show an image of a landscape to an array of ecologists and if each individual were asked privately what relevant ecological phenomenon was portrayed in that image, every response would be different in some respect (Box 2.1).

Differences among colleagues' perceptions are more profound than this exercise demonstrates. They are only a glimpse of how we differ in

Box 2.1
Diversity of perceptions expressed by ecologists

This image of the Snowy Range in the Medicine Bow Mountains of Wyoming (below) was distributed to 15 ecologists at the University of Wyoming with this question: "... what is the first ecological message/concept/paradigm/story/example that comes to your mind in this image?"

The following illustrates the range of responses to this image.

1. Examples of community development following disturbance: rockslides, glaciation.
2. Limits to seedling establishment in this arid, treeline environment.
3. The Snowy Range may be one of the most exemplary (as well as accessible) examples of the subalpine/alpine ecotone in the Intermountain West.
4. The impact of humans on fragile systems.

> ## Box 2.1 (cont.)
>
> 5. Humans like other large, but not small, mammals make paths that contour the landscape with a minimal slope.
> 6. Trails may act as vectors for invasive plants.
> 7. This might be a common scene for August ... but the complete lack of snow seems unusual even for late summer.
> 8. ... a patchy landscape dominated by environmental (particularly climate) extremes, with minimal direct human impact.
> 9. ... heterogeneous landscape [such that] bare rock, forest patches, a lake, and human construction means that the spatial arrangement of these patches will dictate function.
> 10. Human domination of ecosystems.

our fundamental beliefs in what exists – what is real in the universe. We differ in our acceptance of the reality of entities composing the very foundations of ecology: organisms, species, populations, communities, etc. We have staked out different positions on the degree to which we can generalize about nature and the best ways to understand nature in our not-so-common quest for truth(s). In fact, some ecologists believe that such a quest is a quixotic illusion and should be abandoned. Instead, they assert, ecology's main task should be to solve actual problems in particular cases; the search for generalities is but a fool's errand. Beyond conscious beliefs, we also differ at a psychological level in our perceptions as they are translated into pattern recognitions. This view of cognitive behavior may be termed heuristic thinking, a semi-conscious but adaptive mode of rapidly dealing with perceptions and organizing them as cause-and-effect relationships.[1] Heuristic patterns differ among ecologists as well as between ecologists and non-ecologists. The responses in Box 2.1 are realizations of the differences in our heuristics.

Most ecologists would probably agree with the realist's stance: "nature has an objective existence independent of any perception of it, and that observation can disclose the laws governing natural systems".[2] Furthermore, they are likely to adopt naturalism or a naturalistic perspective such that, "there is but one system of reality – nature – and this system does not depend on supernatural factors."[3] But, if pressed, problems quickly arise. For example, what do we mean by "nature?" Quoting Ferré,[4] Keller and Golley[5] say there are two meanings of nature, one is

---

Box 2.2
"Knowing" isn't easy

William Wimsatt addressed the way we think, as well as our wish to know:[1]

We need to be much less absolutist, and much more contingent, contextual, and historicist in our analyses of science. But we must do this by recognizing the real complexities we are increasingly able to study in natural systems whose simplicity we have been taking for granted for decades or centuries. A major fraction of these complexities are not a function of our conceptual schemes, language, or interests, but products of the way the world is. Realism lives! But any wise realist must recognize that the social, cultural, and ideational entities of the "social relativists" are real, too, and have causal effects in this world. They must be imbedded in the appropriate (pan-realist) world picture along with the entities of the natural sciences, via the idealized models one constructs, and the carpentered "natural" entities, tools, practices, procedures, and phenomena one detects and experiments with and on – guided, regulated, and maintained by the social structures, languages, and values of science. All of these are real in our world, and we mold and are molded by and in their presence and effects.

---

that nature is everything apart from the artificial (that is, human-made). The other is that nature connotes everything apart from the supernatural. In fact, ecologists use both meanings in different contexts,[6] and such equivocation suggests both individual and collective confusion of meaning.

Ecologists generally understand that the complexity of our subject precludes complete knowledge. But others contend that even with the best of methods, we can only dimly perceive nature as through a darkened, distorted glass. To them, subjectivity is inescapable, stemming from influences of our ecological schools (or "scholasticism"[7]), geopolitical frameworks, social contexts, and cultural values[8] – as well as being deeply rooted in cognitive levels.[9] These influences prohibit objectivity, such that our science-based understanding of nature is at best one of "mitigated realism".[10] While there is a wide range of beliefs among us as to the degree to which we can know reality, we take the position that, with care, ecologists can understand nature (Box 2.2).[11]

A defining property of any science is the spectrum or array of aspects of nature that it addresses. Ecology covers an enormous range of phenomena. Keller and Golley recognized three disciplinary meanings of ecology itself: "romantic ecology" (perhaps better called aesthetic ecology), "political ecology" (perhaps better termed axiological or values-based ecology), and "scientific ecology."[12] But opinion is scattered as to what constitutes "scientific ecology" (Box 2.3 and associated references in endnote[13]). Robert McIntosh[14] noted:

> It seems that since the rise of ecology to high popularity in the 1960s, almost everyone is prepared to define or delimit ecology and, having done so, to say whence it came and to answer the question frequently raised in the early years of ecology, what good is ecology?

Box 2.3
What is ecology?

McIntosh (1985) has provided a scholarly analysis on the origins of the term and science of ecology.[13] The following are examples illustrating some of the contrasting definitions by which ecologists live and work.

By ecology, we mean the body of knowledge concerning the economy of nature – the investigation of the total relations of the animal both to its organic and to its inorganic environment; including above all, its friendly and inimical relation with those animals and plants with which it comes directly or indirectly into contact – in a word, ecology is the study of all the complex interrelationships referred to by Darwin as the conditions of the struggle for existence. This science of ecology, often inaccurately referred to as 'biology' in the narrow sense, has thus far formed the principle component of what is referred to as 'Natural History.'

> *(Haeckel (1870) (English translation from Allee et al. 1949))*

Note the focus expressed through the restrictive reference to "animal."

Ecology may be defined broadly as the science of the interrelations between living organisms and their environment, including both the physical and the biotic environments, and emphasizing interspecies as well as intraspecies relations.

> *(Allee et al. (1949))*

This is a "modern" definition that places an especially strong emphasis on biotic interaction and, in spite of the title of the book from which it comes (Animal Ecology), *is not animal-centered.*

(Ecology is) the study of the interrelationships of organisms to one another and to the environment.

> *(Hanson (1962, p. 121))*

Box 2.3 (cont.)

This is the standard definition, versions of which are repeated in Colinvaux (1973, p. 3), Owen (1974), Cronan (1996), and Collin (2004).

> it is more in keeping with the modern emphasis to define ecology as
> the study of the structure and function of nature [his italics underlined].
> It should be thoroughly understood that mankind is part of nature,
> since we are using the word nature to include the living world.
>
> *(E. P. Odum (1963, p. 3) (after the standard definition))*

While attractive in some ways, this definition is really indistinguishable from science in general.

> The subject of ecology is the relation of living organisms including
> human beings to their environment.
>
> *(Peters (1991, p. xi))*

Later, Peters says that areas of predictability like lake phosphorus concentrations are not ecology; ecology must be focused on organisms to be ecology. Thus, ecologists don't even agree on what "slice of nature" we purportedly study.

> [E]cology is the scientific study of the interactions that determine the
> distribution and abundance of organisms.
>
> *(Krebs (2001))*

Krebs' definition, while narrow for some, is highly considered. In his view Haeckel (1869) is too broad to be useful; Elton (1927) is uncomfortably vague; Odum (1963) is not completely clear, and Andrewartha (1961) ignores relationships. At least Krebs' definition restricts ecology to a distinguishable subunit of science. However, his restrictions ignore large sectors that other self-declared ecologists might include such as collective levels of organization (e.g. communities, ecosystems, and landscapes), as well as biologically driven energetic and biogeochemical phenomena.
Do not have a definite definition of their own but say,

> Broadly speaking scientific ecologists tend to have two definitions of their
> subject, each of which captures something different about what we mean
> by ecology. The first definition is that ecology is concerned with the
> interaction between organisms and the environment. The second stresses
> that ecologists are trying to understand the distribution and abundance
> of organisms.
>
> *(Cotgreave and Forseth (2002, p. 2))*

Pickett *et al.*[15] have admirably classified the range of definitions of ecology as paradigmatic or viewpoint positions. Their taxonomy, listed below, simultaneously synthesizes the range and illustrates the disparity of understandings of ecological science among dedicated ecologists.

1. *Ecosystem paradigm*: The study of the structure and function of nature.[16]
2. *Population paradigm*: The study of the interactions that determine the distribution and abundance of organisms[17]; the study of the natural environment, particularly the interrelationships between organisms and their surroundings.[18]
3. *Toward integration – organism centered*: The scientific study of the processes influencing the distribution and abundance of organisms, the interactions among organisms, and the interactions between organisms and the transformation and flux of energy and matter (Institute of Ecosystem Studies definition).[19]
4. *Toward integration – general*: The study of ecological systems, and their relationship with each other and with their environment, where "ecological system" is defined as any natural or arbitrary unit at or above the organismal level of complexity.

Even if all ecologists could comfortably find themselves in these four classes of definition, the range of ways they see nature is obviously broad. Different perceptions derive from, and give rise to, different perspectives on what is real, what are the properties of those things deemed to be real, what is worth asking, what constitutes knowledge, and what methods are valid in learning about the natural world. The next section explores how ecologists differ not only in their perceptions of nature, but in their approaches to learning about the natural world.

From the roots to the twigs of ecology: uncertainty about what is

Much of what is written about science philosophy centers on the ways we claim to know what is true and how we go about gaining this knowledge – epistemology.[20] Less attention has been devoted to the fundamental commitments we make to things that we think exist and their properties. Keller and Golley note that, "Surprisingly little work has been done on . . . the metaphysical character of ecological entities and processes . . . in the philosophy of ecology."[21] Consequently, they open their excellent book with a discussion of entities, objects or things that are the foci of study in ecology. We wholeheartedly agree with their observation of ecology in this sense. We hold the view that an ontological commitment – although often implicit and unexamined – to the existence of particular entities and processes is central to an ecologist's philosophy.[22] And the kinds of properties that one holds as being fundamental to these entities and processes constitute the metaphysics of one's science. It appears to us that much of ecology's uncertainty stems from ontology and metaphysics, while most of the debate pertains to epistemology.

Ecology's broad domain of the interactions of life and the abiotic environments on Earth leads to identification of the entities it recognizes as relevant to science, and at least operationally real.[23] These are commonly understood to be: genes, individual organisms, species, niches, populations, guilds, communities, ecosystems, landscapes, and the biosphere (Box 2.3). But in what sense are these entities "real?" Are they tangible in the sense of levers and planets,[24] or are they abstractions designed to help us think about natural systems?[25] Consider that even the most widely accepted entity – the individual – is frustratingly vague. In spite of the powerful intuition humans have about the reality of organisms,[26] not all members of global biota that we might define as individuals are unambiguously discrete: e.g. mutualistic protists, clonal plants, lichens, fungal and coral colonies, social insects, etc. (Box 2.4).

Species are no better understood among ecological entities. The underlying evolutionary character of ecology forces us to acknowledge species as idealized (but convenient, often useful, and sometimes necessary) abstractions, with no properties that can be generalized across all life forms (Box 2.4).[27] Likewise, ecologists are keenly aware of the assumptions implicit in the meaning and boundaries of "populations." Populations are only what we say they are for particular purposes. The same is also true for "guilds," "feeding groups," and "niches." The last is an especially clear case of our recognition of multiple meanings, each of which can be useful to ecologists.[28]

While we lack clarity in our understanding of what is meant by a species or population, vagueness deepens with respect to the reality and properties of "community." The philosophical battles over coherent, integrated communities versus stochastic, individualistic communities have been fought for nearly a century.[29] To a large extent, opponents have been talking past one another without clarifying their conceptual

Box 2.4
When the species concept breaks down

Biologists have long conceded that the designation of species identities is, to a large extent, a matter of judgment and specialists' opinions. There are criteria, to be sure, but the criteria for differentiation vary and tests of reproductive isolation may be difficult or impossible to conduct in many cases and circumstances. The incorporation of molecular techniques into phylogenetic systematics demonstrates genetic differences, and thus, ancestor-descendant relationships, at a more fundamental level (but see Box 2.3). This has led to extensive revisions of morphological- and cytogenetic-based phylogenies, but this approach doesn't necessarily confirm the validity of designated species.

The species concept may be under the greatest attack among biologists specializing on prokaryotes. Because prokaryotes have simpler genomes and are more amenable to molecular analysis, they offer richer opportunities for genetic understanding than do most eukaryotes. Such research has illustrated extensive lateral gene transfer among prokaryotes and evidence of extensive pangenomes among "species" in natural, as well as laboratory, environments. In fact, a lively debate on the microbial species concept is ongoing among prokaryote biologists, with some proposing that the species notion be given up entirely and that functional terms be substituted instead (e.g. operational taxonomic units, molecular clusters, or ecotypes; for in-depth information, see endnote[22]).

structures or even recognizing them as such. An early statement by Stanley Cain[30] illustrates the confusion about the reality of "community":

> No ecologist doubts the existence of individual, socially integrated communities – the stands, the association-individuals, the biocoenoses – in their concrete expression in a certain place and at a certain time . . . The association in the abstract, or community of any rank is another problem . . . Science requires the formation of concepts and scientific thought attempts clarification of these.

In fact, Cain's confidence in the "existence" of concrete "biocoenoses" or "stands" is as misplaced as the belief that natural kinds of communities have objective existence beyond their definition by scientists. The onto-logical problem only gets worse as we consider increasingly inclusive (and abstract) entities such as community types and formations. One could argue that "community" should not be regarded as an entity at all, in either a collective or local (concrete) sense. It probably never has been used in the collective sense of all types of organisms, but only in a restrictive way for taxonomic assemblages of certain groups such as birds and plants. Thus, "community" is not even semantically correct. "Assemblage" may be a more accurate term for what we have meant by "community." Otherwise, "community" might be a useful abstraction for organizing observations of nature – a kind of practical framework (*sensu* Pickett *et al.*).[31]

"Ecosystem" was carefully defined by Arthur Tansley,[32] but the term went through some drastic transformations by subsequent users,[33] such that it came to inherit some of the same problems of "community."[34] These problems included teleological and emergent properties – as opposed to reducible or aggregative[35] – along with arguments across the generations of ecologists about reductionist versus holistic approaches. Pickett and Cadenasso[36] revealed three distinct dimensions of the term "ecosystem": meaning, model, and metaphor. It is uncertain whether ecologists themselves are clear in their own minds which dimension they are addressing in given situations. Particular points of view can be pulled from the literature (admittedly out of context) to dramatize the variety of understandings of "ecosystem". For example, in an aptly titled paper, "Do Ecosystems Exist?" Carl Jordan asserted:[37]

> If ecosystems are tautologies, ecosystem study belongs in departments of philosophy or religion, since tautologies are not amenable to scientific testing. In contrast, if ecosystems are more than definitions, i.e., they are real-world information networks, then they are amenable to scientific tests and are proper subjects of scientific investigations.

This quote illustrates three points. First, ecologists are not always con-scious that all ecological entities entail some ontological account, includ-ing the possibility that they are products of definition and that their metaphysical properties depend on criteria that are themselves matters of contention. This point is exacerbated further in this quote by the vague phrase, "real-world information networks." Such naïve realism assumes that the objects posited by scientists actually exist and fails to ask either what is meant by "information networks" or how we know these things exist in the "real-world." Second, this paragraph misuses "tautology" – a statement which is necessarily true by virtue of its logical form."[38]

This is a question of ontology, not logic. Third, this quote points to a lack of understanding of the relationship between science and philosophy (as if these two disciplines were entirely independent; see Box 2.5) and religion and philosophy (as if one could not discern what sorts of claims were within the purview of each discipline).[39]

Ecosystem is an abstract ontology that has value for particular purposes (the same might be said of numbers, which are no less real for being useful abstractions). Regrettably, ecologists struggle with an allegedly tangible reality while others see it for what it is, as illustrated by this quote from Loehle and Pechmann:[40]

Box 2.5
The relationship of science and philosophy

Although some ecologists would consider their science either a subset of philosophy or a largely independent enterprise, we would maintain that the relationship is rather more complicated than shown in either of these Venn diagrams:

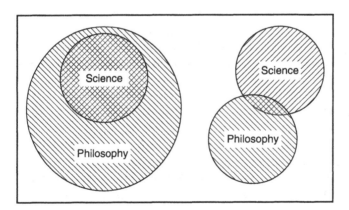

While it is the case that science entails philosophy, there are elements of science that are more effectively understood in terms of psychology, sociology, history, or technology. We would suggest that philosophy is something akin to the *Constitution* of science – a set of broad principles and guidelines which constrain the nature of the discipline but do not prescribe specific, and often important, actions

---

Box 2.5 (cont.)

and decisions. And, as with the Constitution, the philosophy of science is a "living document" that must be interpreted and can be amended, although such changes are not made in a cavalier manner.

As such, the ecologist's laboratory or field site is analogous to a crime scene and the journal or grant review process is similar to a courtroom. In each setting, a practitioner is not allowed to violate the fundamental understandings of the community without risking sanction (e.g. the lawyer cannot lie to the court and the ecologist cannot rely on miracles to explain the data). However, there is a very great deal of latitude within these broad limits, such that states and municipalities can differ dramatically in their laws and judges can vary greatly in their sentencing, just as ecosystem and population ecologists can differ dramatically in their theories and journal editors can vary greatly in their criteria for publication. We might further note that civil disobedience and jury nullification have their parallels in the scientific community, but these are controversial practices, as within the legal realm.

Some do not believe that ecosystems are actual entities. They hold that ecosystems are no more than a collection of species in a particular environment, with all interesting questions centering on species' adaptations.

If established ecologists are confused about the reality of the entities we study, we can only imagine the uncertainty of students as they leave the ecology classroom.

"Landscape" is perhaps the most ill-defined concept of all for ecologists due to its semantic and dual origins in human affairs and ecology.[41] Further, this unfortunate term is etymologically confounded by its terrestrial, folk language origin, preventing easy adaptation to marine littoral, lentic or lotic settings. Yet, while landscape ecologists consciously live with considerable ambiguity and differences of opinion about the "landscape" concept,[42] they make very important contributions to our understanding of nature.

The largest ecological entity – biosphere – would seem to have the greatest potential for having a single meaning for ecologists. Surely the planet is as discrete an entity as we could imagine. What could be more tangible than the "biosphere?" However, we find that biosphere means different things to different people. Earth scientists originated the concept "biosphere" as the total mass of living and (and sometimes, dead) organic materials. Thus, the biosphere is not a place, but a mass, in spite of the suggestive suffix, "sphere."[43] Some ecologists, on the other hand, envision the biosphere as a shell-like spherical space in which living things occur[44] or in which ecosystems function.[45] By either meaning, biosphere is relatively tangible compared with species, guild, niche, population, community, ecosystem, or landscape. Nevertheless, one cannot assume what is meant by a particular individual without further clarification.

Properly understood, ambiguity can spawn creative pluralism. The existentialist philosopher Simone de Beauvoir noted in *The Ethics of Ambiguity* that: "To declare that existence is absurd is to deny that it can ever be given a meaning; to say that it is ambiguous is to assert that its meaning is never fixed, that it must be constantly won."[46] Likewise, the pragmatic philosopher, Charles Peirce[47] argued that there is, or can be, virtue in vagueness, to the extent that it allows a concept to be useful. A tool that can do exactly one thing can be limited to the point of near uselessness (e.g. a hammer that drives only 16-penny galvanized nails). Alfred North Whitehead went so far as to suggest that clarity and importance were inversely related.[48]

The scholarly labor of continuing to explore, refine, revise and otherwise make sense of ecological concepts such as landscape, ecosystem, community, and species – terms which continue to engender intellectual discourse and constructive discord – is a vital element of a living scientific discipline. Once dissent is suppressed and the matter is settled, these terms no longer have the capacity to stimulate inquiry. Indeed, given the importance of skepticism, the reification of concepts could well be taken to be inimical to the scientific venture. As such, it becomes the professional duty of individual ecologists to sustain a vibrant discourse and resist the social pressure towards easy consensus. However, this does not justify impeding the majority in the name of respecting a minority view that has been authentically heard, carefully considered, and fairly rejected – unless or until new evidence or arguments are available. In the same sense, philosophy's perennial struggle to define truth, beauty and goodness should be taken not as evidence of an intellectually weak enterprise but as an indication of having identified important, difficult, and compelling problems.

The uncertain nature of ecological entities reflects both human interests and objective existence.[49] While such ambiguity can allow productive discourse, a lack of clarity in other terms used by ecologists – those pertaining to the science of ecology, rather than the subjects of its study – has generated rather consistently negative consequences. The next section of this chapter moves from an analysis of the philosophical basis of the entities we study to the problems with the way ecologists describe the nature of these studies in philosophical terms.

The tangled terminological bank

As we discussed how ecologists struggle with the nature and reality of entities such as "species" or "gene," we now begin to explore the confused ways in which ecologists use terms such as "theory" or "law." We do not mean to suggest that our difficulties are unique in this regard. Indeed, we can argue that practitioners in genetics and systematics are also unclear as to the meaning of "gene" and "species." Likewise, it may well be the case that chemists and physicists wrestle with what "theory" and "law" may be in a manner similar to that of ecologists. But the authors' knowledge does not extend into these other realms. Indeed, claiming to have expertise in a field as broad and deep as ecology is audacious. If readers from other disciplines, such as genetics or chemistry, find our analysis apropos to their fields, then this will be most welcomed. We strongly suspect that ecology is not unique, but without authentic experience in other sciences, we do not presume to know the full nature of the similarities and differences.

Ecologists not only engage different entities and processes, but they also inhabit different scientific cultures, observe nature through different lenses,[50] ask questions at different scales, and employ different methods.[51] For these reasons, terminological inconsistencies are commonplace even between seemingly closely related groups of ecologists. In the dialogues between the cultures and subcultures, words have different meanings. Examples of troublesome nouns include: notion, assumption, axiom, fact, generalization, understanding, explanation, confirmed generalization, concept, hypothesis, theory, principle, general theory, contingency theory, law, model, paradigm, framework, translation mode, hierarchy, scale, and domain.

Most of us probably have our own working definitions of what these terms mean, but dare we expect common ground on these terms with even our closest colleagues? Our review of a broad literature suggests that the answer is a resounding "no." Consider the widespread confusion about the commonly and variously used term "paradigm." A 2005 edited book titled, *Ecological Paradigms Lost: Routes of Theory Change*[52] illustrates the paucity of discussion directed toward, much less the agreement on, the meaning of "paradigm" itself. Such discussion as appeared was vague and dismissive, boldly highlighted in the book's Foreword, where Robert Paine admitted, "The term *paradigm* has stuck; but like *niche* and *community*, it defies strict, generally acceptable definition ... Ecology seems to have a surfeit of paradigms."[53] Some contributing authors credited Kuhn[54] with the contemporary meaning of paradigm as a kind of "normal science" while others continued to use paradigm as a synonym for a theory or a concept. In his review of this book, Craine[55] concluded that:

> The question that needs to be asked first is not whether ecological paradigms shift, but what the paradigms are. It is clear that paradigms in this field are hard to define. Are there really (or were there ever) competing paradigms in ecology, or are there just a lot of people who believe different things?

Under these circumstances, it could be argued that paradigm serves no useful purpose in ecology unless it is defined by the user for every case. In other words, it is a classic "non-operational construct – concept cluster, conflation, pseudocognate, omnibus term, penchestron, non-concept" of the type Peters assailed as a barrier to the advancement of ecological science.[56]

Another central term with different meanings for ecologists is "theory." Paine's candid description of his perplexity about this word is an honest expression that many of us can share, but perhaps few would admit to our students or peers.[57]

> What exactly constitutes a theory? Dictionaries are of little help because they embrace wording from qualitative "contemplation" to "scientifically acceptable general principles offered to explain phenomena" ... MacArthur (1968) caught the term's sense as employed by most ecologists: "A theory must be falsifiable to be useful to a scientist, but it does not in itself have to be directly and easily verified by measurement. Most often it is the consequences of the theory that are verified or proved false." Hypotheses identify testable cause-and-effect mechanisms. Thus cosmologists continue their quest for a grand unified theory coupling gravity to other forces; that theory is a goal, approached by testing predictions or incorporating novel phenomena. But does ecology have a dynamic and quantitative theory of succession, arguably ecology's first organizing principle, or food web organization, or biodiversity, or ecosystem functioning? And if modeling has been applied to these verbal cornerstones of ecology, do they generate testable predictions?

To further support the nature of Paine's perplexity, a range of meanings for theory is presented in Box 2.6.[58]

Box 2.6
Theories about theory

This sequence of statements is designed to extend from the most defined and discrete meaning of "theory" to more generalized concepts of the term. Citations are provided in endnote[58].

From the Concise Routledge Encyclopedia of Philosophy: a scientific theory is "a conceptual device for systematically characterizing the state-transition behavior of systems."

A specifically operational definition of theory advocated by Peters states, "all constructs that make potentially falsifiable predictions are called 'theories.'" *Thus, predictability would be essential to separate a notion or concept from a theory.*

Hilborn and Mangel say that theory is: "a systematic statement of principles involved or 'a formulation of apparent relationships or underlying principles of certain observed phenomena which has been verified to some degree.'" *They differentiate hypothesis from theory with the former being* "an unproved theory, proposition, supposition, etc., tentatively accepted to explain certain facts or to provide a basis for further investigation."

Pickett et al. *take a broader view of theory. They consider theory to be a* "system of conceptual constructs that organizes and explains the observable phenomena in a stated domain of interest that captures most if not all the attributes of theory accepted by the modern view." *They contend that such systems consist of notions, assumptions, concepts, definitions, facts, confirmed generalizations, laws, models, generalizations, theorems, translation modes, and hypotheses – all of which demand their own definitions.*

Ford, like Pickett et al., *takes a broad view as well. He writes,* "a theory has two parts. The working part of a theory is represented as a logical construction comprising propositions, some of which contain established information (axioms) while others define questions (postulates) ... The working part of a theory provides the information and logical basis for making generalizations."

The fact that ecologists do not have a shared understanding of terms as central as "theory" arises from our lack of a shared understanding of the meaning, methods, goals, and limits of science in general, as well as within our science in particular. This lack of a common philosophy is mirrored and caused by divisions among ecologists working at particular levels of focus (e.g. species, habitat, community, or ecosystem). In fact, even among papers on the philosophy of ecological science, analyses are almost always presented in the context of a single level of organization or focus, as if population ecologists had no relevance to behavioral ecologists or a framework for studying communities could be developed without regard to ecosystems.[59] Consequently, communication and understanding is limited within ecology, not to mention in its relationship with other sciences.

Branching ways of knowing

To this point, we have focused on confusion and uncertainty with regard to the nature of ecology and of the entities and processes that we study. Within the philosophical realm of seeking truth, these topics fall under the categories of ontology and metaphysics (see Glossary). We agree with Keller and Golley[60] that these areas of philosophy have received much less attention than determining how we know what we know – that part of philosophy termed epistemology. Epistemology addresses knowledge, typically taken to mean having a justified true belief, although there are interesting problems with this basic account (e.g. true beliefs whose justifications appear sound but do not accord with the world). That is, we only know something if we can provide some justification for what we believe (being accidentally correct does not constitute knowing) and if what we believe is true (we cannot genuinely know something that is false, and history shows that science has been rife with justified untrue beliefs).

In science, the justification of belief takes a surprising range of forms, some widely recognized, debated or accepted, but some others relatively cryptic and insidious. Even among those forms that are relatively well-known, there is debate about their meaning and usefulness – often following dichotomous arguments. These dichotomies, illustrated below, are partially due to semantic issues, but many spring from very real differences in ecologists' ways of understanding nature. These are not necessarily false dichotomies, although it may often be the case that they are cast as false dilemmas such that alternative branches, a plausible continuum, and middle positions are precluded. Thus, if understood properly, the dichotomies may be useful in representing ends of a valid spectrum. Otherwise, they pose obstacles for understanding our own science.

Rationalism versus empiricism

Perhaps the most apparent dichotomy has been between those who develop "general" theories via deduction from general, fundamental biophysical properties (rationalists), as opposed to those who emphasize the development of theory through observation of the world (empiricists). One way of understanding the difference was suggested by William James[61] who asserted that rationalists (he used the term "scholastics") pin their claim on the principle from which a concept originates while empiricists rely on the outcome or result of an idea.

Rationalists hold that because of nature's bewildering diversity, induction from empirical evidence cannot reveal the general underlying patterns. Instead, pervasive first principles of how ecological nature works must be deduced either through logical-mathematical processes (e.g. competition for finite resources by exponentially growing populations) or from biophysical axioms.

In contrast, "empiricist proclivities run strong in ecology"[62] although there is much diversity, even conflict, in the empirical methods. Keller and Golley[63] are correct in saying that empirical ecology is variable (perhaps to the point of incoherence) in its approach. Experimental methods are often applied in a fragmented way as described by Schrader-Frechette and McCoy,[64] sometimes under the guise of a superficial version of pluralism.[65] Debates about the merits of rationalism versus empiricism characterize much of the science philosophy literature of ecology.[66]

Naturalism/realism versus nominalism

Orthogonal to the dichotomy of rationalism and empiricism, is the contrast between naturalism (or realism) and nominalism. Realists maintain that entities and phenomena can be grouped into sets or natural kinds, and these correspond to actual classes in the world, allowing us to understand and describe nature with relatively simple and pervasive theories or laws.

Nominalists are committed to the view that all of nature is unique and each case is particular. As such, all groupings are human contrivances that fail to reflect objective reality. This dichotomy is exemplified by the early debates on the nature of community (Clements versus Gleason[67]) and lives on today through realist perspectives implicit in some forms of "theoretical ecology"[68] versus nominalist viewpoints from the empirical and applied camps in ecology.[69]

Our brief descriptions oversimplify these foundational, philosophical dichotomies. Although the accounts we offer provide a sense of the critical contrasts, differences between the two points of view are heavily nuanced. There are, for example, numerous subcategories of both realists and nominalists, some of which will be described later when they are relevant to developing a philosophy of ecology.

Meta-methodological dichotomies

Two other dichotomies aggravate debates among ecologists. The first is reductionism (the position that we can understand the world through a complete grasp of its component parts – that all biological facts are explainable in terms of physics) versus holism (the position that the world can only be understood in terms of networks of relationships that are not reducible, but exhibit emergent or supervenient properties). The second source of conflict is determinism (the position that every event is necessitated by antecedent events and conditions together with the laws of nature) versus stochasticism (the position that events may be influenced by randomness or chance, which is not contained within preceding events or necessitated by the laws of nature). We regard these as "meta-methods" within the realm of epistemology. Although these issues are philosophically rich and relevant,[70] we will not pursue them in depth here. Rather, we will more fully develop the dichotomous nature and consequences of these concepts as we explore their particular relevance to ecology.

What is true? What is belief? What is justification? And, how do these interact to provide knowledge? Epistemology might be argued as constituting the most important aspect of philosophy for science and certainly has received the most attention in ecology. For many ecologists, however, epistemology *is* science philosophy. The acceptance of multiple ways of knowing and justifying that knowledge has been presented as "pluralism," and some ecologists have advocated this as an approach to resolving, or avoiding, debates about epistemology.[71]

We believe that there is a place for epistemological pluralism as long as it is more than just blind tolerance for the sake of quashing argument. That is, it is important for ecologists to discriminate and reconcile the instances in which each of the various approaches to knowing may be useful and appropriate. There would be considerable resistance to accepting a hypothesis because it reflected a political view, funding a proposal because the experiments struck us as beautiful, throwing out data based on Ouija board readings, or explaining trends as the result of divine intervention – but why? We contend, as well, that beliefs about epistemology are linked to beliefs about ontology and metaphysics (and arguably ethics, at least as a constraint to the ways in which we pursue knowledge) so that those areas of philosophy must become familiar as well. Authentic pluralism requires a larger measure of philosophical understanding than just epistemology.

Our cacophonous confusion

We assert that conceptual and methodological confusion in ecology is pervasive within and among individuals, and multidimensional in character. We ecologists are unclear as to the boundaries of nature, the nature of reality, the meaning of philosophical terms, and the existence of the entities we study. At the same time we are committed to various ways of knowing and degrees of belief in the generality of what we claim to know. Does this matter? What are the consequences of this cacophony?

If the philosophical disparities were just a matter of our practicing multiple approaches to understanding ecology, then variety might create a richer understanding overall – creative pluralism, rather than destructive incoherence (for a recent example, see Box 2.7; essential references are in endnote[72]).

While the potential for constructive conflict may exist, we are more concerned with the actual disadvantages of discord among ecologists. If we do not appreciate the virtues of diverse perspectives or understand one another well enough to listen to alternative points of view, we will continue to pour valuable intellectual resources into battles that are unwinnable. It may often be the case that opposing positions are contextually valid and yet incommensurable, irreducible and unresolvable, in terms of a unified ecology. They are incommensurable, in that there may be true or useful understandings of the world (e.g. density-dependent and -independent regulation of populations) that cannot both yield the same outcome if acted upon in a particular instance. They are irreducible in that opposing insights are not reducible to any common principle or foundation, as, for example, the concepts of logistic growth or keystone species. By irresolvable we mean there are no rational (in the common usage sense) means by which we can settle or determine the optimal trade-off or balance between competing understandings. In such cases, we must reach conclusions through means other than pure reason (e.g. sociopolitical

---

Box 2.7
Attempting to resolve a contemporary theory

Ecologists have responded to the promulgation of the "Metabolic Theory of Ecology" with strongly held opinions. Recently Martinez del Rio sought an explanation of disparate attitudes to the theory. He introduces his objectives with the following argument:

The metabolic theory of ecology (MTE) is bold, big and contentious. It has been heralded by some as one of the greatest advances in ecology, and by others as a threat that will slow down conceptual progress ... A curious observation on the published critiques of MTE is that most include a sentence in which the authors applaud how the theory has sparked controversy and focused attention on a potentially important issue. I am less sanguine about the presumed benefits of heated debate, inasmuch as these are based on the assumption that scientific controversies are resolved rationally when logic and/or evidence hold sway. Scientific controversies are about facts, observations and methods, but they are also about complex implicit assumptions about how nature works and about how it should be studied ...

Martinez del Rio seems to say that rationalism has marginal value in this debate – that empiricism holds the keys to confirmation. At the same time, he does place value on logic, although precisely what he means by logic holding sway is not clear. We take this to mean reasoned arguments about empirical evidence, but it could also refer to models based on first principles which Peters (1991, p. 18) would include as part of tautological processes that others might term rationalism. Later Martinez del Rio says:

Although my primary objective is to diagnose why MTE is so controversial among biologists, this paper also has a prescriptive element. It will contend that progress in the development of MTE will be faster and less contentious if we adopt a model-centric, rather than a theory-centric, perspective.

We interpret Martinez del Rio's solution to emphasize the empirical content and support for the theory rather than the rationalist basis that was the ostensible origin of the theory. Thus, this would seem to be a contemporary manifestation of the rationalist versus empiricist debate. At the least, he is attempting a positive, philosophically based effort at reconciliation signaling a pluralist approach to ecology. Incongruously, though, he then makes the following statement which seems contradictory

---

Box 2.7 (cont.)

to his contention that the heated and fruitless scientific debate regarding MTE has been perpetuated by ecologists who have failed to grasp the philosophical foundations implicit in their positions:

> I do not imply that scientists have taken sides in the battles over science philosophy ... Most biologists pay little attention to debates in philosophy, and we are, for the most part, none the worse for it.

It appears that too much attention to philosophy is bad form among biologists. If this sort of anti-philosophical attitude pervades ecology – a possibility that generally accords with our experience but would require a serious, sociological study to establish – then Martinez del Rio may have been carefully sustaining his credibility as a "real" scientist by framing his critique as a rare instance in which philosophical analysis was valid. If so, then ecology might be in rather greater philosophical peril than one would otherwise assume.

Box 2.8
Illuminating forest fires science and policy

The large, uncontrolled and uncontrollable fires of 1988 in Yellowstone National Park stimulated discordant points of view on both the science and management of fires in public, forested ecosystems (Christensen *et al.*, Bonnicksen, and Schullery[73]). Scientific and managerial controversies were overridden, however, by the kinds of debates that society practices – arguments driven by economic, aesthetic, cultural, and emotional responses, in which scientific information and opinion play a relatively minor role.

As the ashes cooled, noisy dissension arose between ecologists (fire and forest ecologists, as well as wildlife biologists) and non-scientific interests, including tourists, purveyors of tourist services, and congressional representatives and the media itself (Smith[73]). The debate over fire management in the West continues to this day, complicated by differing conditions in different parts of the West, and climate change itself (Dale *et al.* and Westerling *et al.*[73]). The socioeconomics of second-home safety and deriving timber products (as opposed to ecosystem services and species diversity)

Box 2.8 (cont.)

from public lands and the actual costs of management are directly connected to political pressures.

In recent years, federal officials have advocated a single policy (Healthy Forests Restoration Act) that may have been too monolithic in promulgation and execution (Stephens and Ruth[73]). There is a science of fire in ecosystems, but that is only a minor part of how truth is sought by individuals and how society addresses such a phenomenon. We would note that more science does not always make for less ambiguous policies and surprisingly, can even make environmental controversies worse (Sarewitz[73]).

---

traditions, moral considerations, aesthetic virtues, and emotional procliv-ities), as might be exemplified with the policies reflecting short-term and long-term benefits of fire in forest ecosystems (Box 2.8 and associated references in endnote[73]).

Science is a communal process. Exposure of ideas for critical review and empirical testing by others sets science apart from other ways of knowing. If we do not bother to listen to others, or categorically deny the relevance of those using different methods, then we close off the essential process of critical peer review regarding our visions of nature. Is this the lesson we want to pass on to our students – that success in scientific enterprise is a matter of building highly constrained careers following the same narrow track that yields more safe publications and thus, funded proposals?

In the following chapter we explore the causes of ecology's conceptual confusion relative to a mythical view of classical physics – the scientific framework which appears to be the ideal toward which many ecologists aspire. Then, from this perspective, we propose how the properties of the slice of nature that we study can be embodied in a philosophical framework embracing the variety of problems and practices that constitute our science.

Endnotes

1. Kahneman, D., P. Slovic and A. Tversky. 1982. *Decisions Under Uncertainty: Heuristics and Biases*. London: Cambridge University Press; Kunda, Z. 1990. The case for motivated reasoning. *Psychology Bulletin*, **108**: 480–498; Wimsatt, W. C. 2007. *Re-engineering Philosophy for Limited Beings: Piecewise Approximations to Reality*. Cambridge, MA: Harvard University Press.
2. Keller, D. R. and F. B. Golley. 2000. *The Philosophy of Ecology: From Science to Synthesis*. Athens, GA: University of Georgia Press, 1.

3. Ibid, 12.
4. Ferré, F. 1996. *Being and Value: Toward a Constructive Postmodern Metaphysics*. NetLibrary, Inc. Publication Info: Albany: State University of New York Press. Source for quote in Keller and Golley, 2000.
5. Keller and Golley, *The Philosophy of Ecology*, 11.
6. Ibid, 12.
7. Pickett, S. T. A., J. Kolasa and C. G. Jones. 2007. *Ecological Understanding: The Nature of Theory, the Theory of Nature*. 2nd edn. San Diego, CA: Academic Press, 172.
8. de Laplante, K. 2005. Is ecosystem management a postmodern science? In *Ecological Paradigms Lost: Routes of Theory Change*, ed. K. Cuddington and B. Beisner. Boston: Elsevier Academic Press, 397–416.
9. Loehle, C. 1987. Hypothesis testing in ecology: Psychological aspects and the importance of theory maturation. *The Quarterly Review of Biology*, **62**: 397–409; Kunda, The case for motivated reasoning.
10. Keller and Golley, *The Philosophy of Ecology*, 12.
11. Wimsatt, *Re-engineering Philosophy*, 148.
12. Keller and Golley, *The Philosophy of Ecology*, 5.
13. Allee, W. C., A. E. Emerson, O. Park, T. Park, and K. P. Schmidt. 1949. *Principles of Animal Ecology*. Philadelphia: W. B. Saunders Co; Hanson, H. C. 1962. *Dictionary of Ecology*. Washington, D.C.: Philosophical Library, Inc; Colinvaux, P. A. 1973. *Introduction to Ecology*. New York: John Wiley & Sons; Owen, D. F. 1974. *What is Ecology?* London: Oxford University Press; Cronan, C. S. 1996. *Introduction to Ecology and Ecosystems Analysis*. Orono, ME: Shaw-Ferguson Environmental Publications; Collin, P. H. 2004. *Dictionary of Environment and Ecology*. London: Bloomsbury; Odum, E. P. 1963. *Ecology*. New York: Holt, Rinehart and Winston; Peters, R. H. 1991. *A Critique for Ecology*. New York: Cambridge University Press; Krebs, C. J. 2001. *Ecology: The Experimental Analysis of Distribution and Abundance*. San Francisco, CA: Addison Wesley Longman, Inc; Haeckel, E. 1869. Ueber entwickelungsgang und aufgabe der zoologie. In *Gesammelte populäre vortrage aus dem gebiete der entwickelungslehre*. Heft 2. Bonn: Strauss; Elton, C. S. 1927. *Animal Ecology*. New York: The Macmillan Company; Andrewartha, H. G. 1961. *Introduction to the Study of Animal Populations*. Chicago: University of Chicago Press; Cotgreave, P. and I. Forseth. 2002. *Introductory Ecology*. London: Blackwell Science.
14. McIntosh, R. P. 1985. *The Background of Ecology: Concept and Theory*. Cambridge, UK: Cambridge University Press, 9.
15. Pickett, S. T. A., J. Kolasa and C. G. Jones. 2007. *Ecological Understanding: The Nature of Theory, the Theory of Nature*. 2nd edn. San Diego, CA: Academic Press, 12.
16. Odum, E. P. 1971. *Fundamentals of Ecology*. Philadelphia: Saunders.
17. Krebs, *Ecology*.
18. Rickleffs, R. E. 1983. *The Economy of Nature*. New York: Chiron Press, 1.
19. Attribution to G. E. Likens by Pickett *et al. Ecological Understanding*, 12.
20. Keller and Golley, *The Philosophy of Ecology*, xi.
21. Ibid, 21.
22. Jax, K. 2006. Ecological units: definition and application. *The Quarterly Review of Biology*, **81**(3): 237–258. Jax states this even more strongly than we do, saying, "The ambiguity of terms and concepts constitutes an impediment to communication among scientists and to the creation of

a coherent theoretical framework in ecology", 238. Also, "An ontological view of ecological units is, as mentioned above, seldom explicit, but it often plays an influential (though hidden) role in the formulation and application of definitions of ecological units and so should not be ignored.", p. 246. But Jax insists that, "On the basis of different understandings of concepts, an object of study ... may be delimited and handled in multiple ways by different scientists. This may result in productive plurality and scientific competition, in the sense of different research schools or research programs ... But such a productive plurality requires scientists to be conscious of this fact and to explicitly express the meanings of the terms they use", 238. Jax discusses populations, communities and ecosystems from several points of view, but appears to take the position that ontological definitions are restricted to those entities that we accept as abstractions, not as tangible objects. We believe that all of our defined objects fall somewhere between highly tangible and total abstractions, and that all are ontologies.

23. Wilkinson, D. M. 2006. *Fundamental Processes in Ecology. An Earth System Approach*. Oxford, UK: Oxford University Press, 4; Haila, Y. 1996. Biodiversity and the divide between culture and nature. *Biodiversity and Conservation*, **8**: 165–181; Reiners, W. A. and Driese K. L. 2004. *Transport Processes in Nature. The Propagation of Ecological Influences Through Environmental Space*. Cambridge, UK: Cambridge University Press, 63.

24. For an introduction to robust definitions of the reality of objects, or entities, see Wimsatt, *Re-engineering Philosophy*, 60.

25. Jax, Ecological units: definition and application, 237–258.

26. When we questioned students as to the reality of what they studied, they were most confident in the reality of individual organisms. They admitted to the arbitrariness of membership of that organism within a species, and the pure convenience of a scientific name, but they were most emphatic that the organism under study was real.

27. For in-depth reviews of the species concept, see: Kitcher, P. 1985. *Species*. Cambridge, MA: MIT Press; Rosenberg, A. 1985. *The Structure of Biological Science*. Cambridge, UK: Cambridge University Press.
For in-depth reading on prokaryotic genomics and considerations of the meaning of species, see the following:
Venter, J. C., K. Remington, J. F. Heidelberg, *et al.* 2004. Environmental genome shotgun sequencing of the Sargasso Sea. *Science*, **304**(5667): 66–74; American Academy of Microbiology. 2006. *Reconciling Microbial Systematics and Genomics*. Washington, D.C., 12 pp.; Tamames, J. and A. Moya. 2008. Estimating the extent of horizontal gene transfer in metagenomic sequences. *BMC Genomics*, **24**(9): 136; Tettelin, H., V. Masignani, M. J. Cieslewicz, *et al.* 2005. Genome analysis of multiple pathogenic isolates of *Streptococcus agalactiae*: implications for the microbial "pan-genome." *Proceedings of the National Academy of Sciences of the United States of America*, **102**(39): 13950–13955; Delaye, L., A. DeLuna, A. Lazcanol and A. Becerral. 2008. The origin of a novel gene through overprinting in *Escherichia coli*. *BMC Evolutionary Biology*, **8**: 31.

28. Peters, R. H. 1991. *A Critique for Ecology*. Cambridge, UK: Cambridge University Press, 91.

29. Keller and Golley, *The Philosophy of Ecology*, 24.

30. Cain, S. A. 1939. The climax and its complexities. *The American Midland Naturalist* **21**: 146–181, 147.

31. Pickett *et al. Ecological Understanding*, 89.
32. Tansley, A. G. 1935. The use and abuse of vegetational concepts and terms. *Ecology*, **16**: 284–307.
33. Bergandi, D. 1995. "Reductionist holism": an oxymoron or a philosophical chimera of Eugene Odum's systems ecology? *Ludus vitalis*, **3**: 145–180.
34. Peters, *A Critique for Ecology*, 91.
35. Wimsatt, W. C. 1997. Aggregativity: reductive heuristics for finding emergence. *Philosophy of Science*, **64**: 372–384.
36. Pickett, S. T. A. and M. L. Cadenasso. 2002. The ecosystem as a multidimensional concept: meaning, model, and metaphor. *Ecosystems*, **5**: 1–10.
37. Jordan, C. F. 1981. Do ecosystems exist? *The American Naturalist*, **118**: 284–287.
38. Peters, *A Critique for Ecology*, 39.
39. Pigliucci, M. 2008. When scientists and philosophers talk to each other. Proceedings of the Stony Brook "SCI_PHI" Symposium. *Quarterly Review of Biology*, **83**(1): 5. According to Pigliucci, "Philosophy is arguably the broadest field of inquiry ever conceived, and it originally included science itself under the rubric of natural philosophy … Science went on to do great things after Descartes, and the two disciplines eventually ended up in different parts of university campuses, often under distinct administrative units. Today, it is unusual to find a scientist with a genuine interest in philosophy, while more and more philosophers feel both that science is relevant to classic philosophical questions and that philosophy, in turn, has much to say to science." We authors take the position that science remains part of philosophy even though scientists may have very narrow understandings of even those areas of philosophy that are relevant to what they do.
40. Loehle, C. and J. H. K. Pechmann. 1988. Evolution: the missing ingredient in systems ecology. *The American Naturalist*, **132**: 884–899.
41. Moss, M. R. 1999. Fostering academic and institutional activities in landscape ecology. In *Issues in Landscape Ecology*, eds. J. A. Wiens and M. R. Moss. Ft. Collins, CO: International Association of Landscape Ecology, 138–144; Wiens, J. A. 1999. Toward a unified landscape ecology. In *Issues in Landscape Ecology*, eds. J. A. Wiens and M. R. Moss. Ft. Collins, CO: International Association for Landscape Ecology, 148–151; Turner, M. G., R. H. Gardner and R. V. O'Neill. 2001. *Landscape Ecology in Theory and Practice. Pattern and Process*. New York: Springer.
42. Reiners and Driese, *Transport Processes in Nature*, 65.
43. Rankama, K. and Th. G. Saham. 1950. *Geochemistry*. Chicago: The University of Chicago Press; Bowen, H. J. M. 1979. *Environmental Chemistry of the Elements*. San Diego, CA: Academic Press; Schlesinger, W. H. 1997. *Biogeochemistry: an Analysis of Global Change*. San Diego, CA: Academic Press.
44. Livingstone, D. A. 1973. The biosphere. In *Carbon and the Biosphere*, eds. G. M. Woodwell and E. V. Pecan. Springfield, VA: Technical Information Center, Office of Information Services, U.S. Atomic Energy Commission, 1–5.
45. Odum, *Fundamentals of Ecology*.
46. de Beauvoir, S. 1948. *The Ethics of Ambiguity*. Secaucus: Citadel Press.
47. Peirce, C. S. 1905. Issues of pragmatism. *The Monist*, **15**: 481–499.
48. Whitehead, A. N. 1929. *Process and Reality: An Essay in Cosmology*. New York: Macmillan. Critical edition by D. R. Griffin and D. W. Sherbourne. 1978. New York: Macmillan.

49. Allen, T. F. H. and T. W. Hoekstra. 1992. *Toward a Unified Ecology.* New York: Columbia University Press.
50. Rozzi, R., J. J. Armesto, B. Goffinet, *et al.* 2008. Changing lenses to assess biodiversity: patterns of species richness in sub-Antarctic plants and implications for global conservation. *Frontiers in Ecology and Environment,* **6**(3): 131–137.
51. Pickett *et al. Ecological Understanding,* 172; Peters, *A Critique for Ecology,* 70, 186; Lakatos, I. 1978. Falsification and the methodology of scientific research programmes. In *The Methodology of Scientific Research Programmes,* ed. J. Worrall and G. Currie. Cambridge, UK: Cambridge University Press, 8–10.
52. Cuddington, K. and B. Beisner, eds. 2005. *Ecological Paradigms Lost: Routes of Theory Change.* Burlington, MA: Elsevier Academic Press.
53. Paine, R. 2005. Forword. In *Ecological Paradigms Lost: Routes of Theory Change,* ed. K. Cuddington and B. Beisner. Burlington, MA: Elsevier Academic Press, xv–xx.
54. Kuhn, T. S. 1970. *The Structure of Scientific Revolutions.* Chicago: University of Chicago Press.
55. Craine, J. 2006. Paradigms undefined. *BioScience* **56**: 447–449.
56. Peters, *A Critique for Ecology,* 74.
57. Paine, Foreword, xv.
58. Craig, E. 2000. *Concise Routledge Encyclopedia of Philosophy.* New York: Routledge (Taylor & Francis Group), 884; Peters, *A Critique for Ecology,* 18; Hilborn, R. and M. Mangel. 1997. *The Ecological Detective: Confronting Models with Data.* Princeton, NJ: Princeton University Press, 24; Pickett *et al. Ecological Understanding,* 63; Ford, E. D. 2000. *Scientific Method for Ecological Research.* New York: Cambridge University Press, 43.
59. Peters, *A Critique for Ecology,* 80–104; McIntosh, R. P. 1987. Pluralism in ecology. *Annual Review of Ecology and Systematics,* **18**: 321–341; Hall, C. A. S. and D. L. DeAngelis. 1984. Models in ecology: paradigms found or paradigms lost? *Bulletin of the Ecological Society of America,* **66**: 339–346. This paper thoughtfully discriminates between cultures and styles of ecology with respect to modeling as a part of theory building; Cuddington and Beisner, *Ecological paradigms,* The search for meaning and nature of paradigms is organized by major sections focusing on populations, epidemiology, and ecosystems.
60. Keller and Golley, *The Philosophy of Ecology,* 133.
61. James, W. 1956 [1897]. *The Will to Believe.* New York: Cosmio.
62. Keller and Golley, *The Philosophy of Ecology,* 134.
63. Ibid, 138.
64. Schrader-Frechette, K. S. and E. D. McCoy. 1993. *Method in Ecology. Strategies for Conservation.* Cambridge, UK: Cambridge University Press.
65. Keller and Golley, *The Philosophy of Ecology,* 139.
66. Chamberlin, T. C. 1965. The method of multiple working hypotheses. *Science,* **148**: 754–759; Popper, K. 1959. *The Logic of Scientific Discovery.* New York: Basic Books; Platt, J. R. 1964. Strong inference. *Science,* **146**: 347–353; Peters, *A Critique for Ecology*; Rigler, F. H. and R. Peters. 1995. *Science and Limnology.* Oldendorf/Luhe: Ecology Institute. Hilborn and Mangel, *The Ecological Detective*; Ford, *Scientific Method*; Pickett *et al.*, *Ecological Understanding.*
67. Keller and Golley, *The Philosophy of Ecology,* 35–54.

68. Watt, K. E. F. 1971. Dynamics of population: a synthesis. In *Dynamics of Populations: A Synthesis*, eds. P. J. den Boer and G. R. Gradwell. Wagningen, The Netherlands: Centre for Agricultural Publishing and Documentation, 568–580; Macfadyen, A. 1975. Some thoughts on the behaviour of ecologists. *Journal of Animal Ecology*, **44**: 351–363; Maurer, B. 1998. Ecological science and statistical paradigms at the threshold. *Science*, **279**: 502–503; Ginzburg, L. and M. Colyvan. 2004. *Ecological Orbits: How Planets Move and Populations Grow*. New York: Oxford University Press.

69. Peters, *A Critique for Ecology*; Schrader-Frechette and McCoy, *Methods in Ecology*; Weiner, J. 1995. On the practice of ecology. *Journal of Ecology*, **83**: 153–158; Rigler and Peters, *Science and Limnology*.

70. Wimsatt, Aggregativity; Schoener, T. W. 1986a. Mechanistic approaches to ecology: a new reductionism? *American Zoologist*, **26**: 81–106; Bergandi, Reductionist holism; Lawton, J. H. 1999. Are there general laws in ecology? *Oikos*, **84**: 177–192.

71. Grene, M. 1985. Perception, interpretation, and the sciences: Toward a new philosophy of biology. In *Evolution at a Crossroads*, eds. D. J. Depew and B. H. Weber. Cambridge, MA: MIT Press, 1–20; MacIntosh, *Pluralism in Ecology*; Keller and Golley, *The Philosophy of Ecology*, 139.

72. Martinez del Rio, C. 2008. Metabolic theory or metabolic models? *Trends in Ecology and Evolution*, **23**(5): 256–260; Whitfield, J. 2004. Ecology's big hot idea. *Public Library of Science, Biology*, **2**: 2023–2027.

73. Christensen, N. L., J. A. Agee, P. F. Brussard, *et al.* 1989. Interpreting the Yellowstone fires of 1988. *BioScience*, **39**: 678–685; Bonnicksen, T. M. 1989. Nature versus man(agement). *Journal of Forestry*, **87**(12): 41–43; Schullery, P. 1989. The fires and fire policy. *BioScience*, **39**: 686–694; Smith, C. 1989. Brave firefighters, endangered national icons and bumbling land managers: network TV myths about the 1988 Yellowstone fires. Paper presented to The Association for Education in Journalism and Mass Communication. 13 August 1989. Washington, D.C.; Dale, V. H., L. A. Joyce, S. Mcnulty, *et al.* 2001. Climate change and forest disturbances. *BioScience*, **51**(9): 723–734; Westerling, A. L., A. Gershunov, T. J. Brown, D. R. Cayan, and M. D. Dettinger. 2003. Climate and wildfire in the western United States. *Bulletin of the American Meteorological Society*, **84**(5): 595–604; Westerling, A. L., H. G. Hidalgo, D. R. Cayan and T. W. Swetnam. 2006. Warming and earlier spring increase western U.S. Forest wildfire activity. *Science*, **313**(5789): 940–943; Sarewitz, D. 2004. How science makes environmental controversies worse. *Environmental Science and Policy*, **7**: 385–403; Stephens, S. L. and L. W. Ruth. 2005. Federal forest-fire policy in the United States. *Ecological Applications*, **15**(2): 532–542.

3

Causes of ecology's conceptual confusion

A philosophical armistice for ecology

Having made a case that ecology is in a state of philosophical civil war, with various, irreconcilable perspectives clashing with one another, we are faced with three options. First, we can simply allow the conflicts to resolve themselves in the hope that the more defensible concepts will emerge victorious. However, such an approach begs the question of whether ecology progresses toward an objective truth – a matter that we will address later. And even if we dismiss the contingency of truths, the history of science does not suggest that such progress is assured.[1] Moreover, the waste of valuable time, intellectual energy, research funds and educational opportunities in fighting such battles (particularly if there is no unconditionally correct perspective) may be deplorable in light of the limited resources available to our science. Our concerns with the "let them fight it out" approach to philosophy are similar to those expressed by other ecologists who have argued that heated debate among scientists is often not resolved – or ultimately even resolvable – via evidence or reason but simmers indefinitely because the disputants fail to acknowledge or address the implicit, philosophical assumptions that fuel the controversy.[2]

A second option would be for us to commit to one or the other side in the various battles and thereby take a stand and perhaps declare victory, but there are two problems with this tactic. There must actually be a right answer (e.g. rationalism is the path to truth, nominalism reflects reality, reductionism is the superior approach), and we must be able to identify this perspective. We are dubious that there is a singular, philosophical view that unconditionally leads to the truths we seek in ecology, and even if such existed, the fact that philosophers and practitioners of science have not yet resolved these matters in fields as old and established as physics suggests that it is most unlikely that ecologists will be able to do so.

The third tactic would be to seek a philosophical armistice. We choose this term intentionally, as it reflects a condition in which the two sides no longer try to defeat one another, without there being a peace treaty. That is, the opponents no longer waste resources in fighting one another, but there remains an unresolved tension – and this seems to be a rather ideal situation for a philosophy of ecology, at least at this point in our history. To understand how such a ceasefire could foster productive, rather than destructive, conflict, it seems necessary to grasp the philosophical foundations that are at work.

In this regard, we will analyze the conceptual differences between ecology and physics, as these disciplines seem to embody much of the underlying antagonism. That is, the philosophical conflicts in ecology are often between two distinct positions. Those who advocate that a form of inquiry is scientific to the extent that it resembles physics battle against those who contend that ecological science has substantially different, but entirely valid, standards for the justification of knowledge. Once these differences are clarified, we'll develop a philosophical approach that sustains the virtue of the intellectual tensions that foster creative work in ecology while putting an end to the open hostilities.

A little comparative theater: idealized physics versus actual ecology

As we have seen, ecology suffers from a philosophical anxiety disorder. Ecologists doubt that theirs is a healthy science – perhaps even a genuine science at all – as it lacks the conceptual clarity and theoretical coherence seen in the paragon of scientific enterprises: physics. Ecologists might well suspect that Lord Kelvin had their science in mind when he said, "In science there is only physics; all the rest is stamp collecting." This perspective is reinforced by the philosophy of science which is largely drawn from, and based on, the physical sciences – primarily classical physics[3] (see Box 3.1).

The sibling rivalry between the physical and life sciences is exacerbated by the former often being cast as the older field of scientific inquiry against which the newcomer is to be evaluated. To be more historically accurate, both sciences are rooted in ancient times and were treated as practical endeavors (e.g. engineering and agriculture) for millennia.

Box 3.1
The philosophy-physics compact

A random sampling of 100 examples used to exemplify philosophical principles in ten different textbooks of science revealed that 86% of the cases were drawn from physics. To further substantiate this intellectual alliance, we conducted a series of web searches (using Google in July 2008) combining "philosophy" and "science" with the following terms:

Search term	# of hits	Ratio
physics	6,120,000	2.5 : 1
ecology	2,430,000	
physicist	44,000,000	19.2 : 1
ecologist	2,290,000	
physical	3,760,000	10.7 : 1
ecological	350,000	

Although ecology arose as a science in the nineteenth century, it seems that the developmental gap came early in the twentieth century with a surge in physics, which ecology failed to match. Perhaps with maturity the philosophical ambiguities in ecology will ameliorate.[4] However, ecological studies have been underway for the better part of 200 years, so the appeal to being a new science seems an excuse rather than an explanation. Rather, we suggest that there are important and fundamental differences between physics and ecology – indeed, between the physical sciences and life sciences, in general, although we'll largely restrict our analysis to physics and ecology. These differences give rise to philosophical disparities such that no amount of time or maturation is likely to turn the pluralistic ambiguities of ecology into the coherent clarity of physics.

While our comparison of ecology to physics will not be unique, there are several considerations that should be addressed before we undertake the analysis. First, others have reflected on the similarities and differences between ecology and physics.[5] In particular, Gasper has argued that:[6]

> This preoccupation with physics is now commonly agreed to have had a distorting effect on the philosophy of science. The tendency has been to assume that certain features of physical theories, such as their tractability to mathematical axiomization, are characteristic of scientific theories in general. To the extent that theories in other areas have not shared these features,

it has been assumed that they are incomplete or deficient and that they need to be developed to fit the model derived from physics.

While such analyses very effectively set the stage with regard to the larger project of reforming the philosophy of science(s), we will seek to frame the comparison in terms that are particularly relevant to our line of inquiry regarding ecology and to identify some (dis)similarities that have not been central to previous analyses.

Second, we would note that we are making comparisons to a particular, and perhaps idiosyncratic, interpretation of physics. We do not claim that this is how contemporary physics actually functions or how working physicists understand their field. Rather, based on our experience and reading, we are presenting physics as it is understood by many ecologists – as an idealized version of classical physics. As such, the formulation of physics that we use is not a straw man, but is our best approximation of how this field appears to ecologists. Our presentation of physics may be a caricature of sorts, but as with a good cartoonist we are attempting to emphasize and highlight those features that attract the attention of the viewer. There are, of course, subdisciplines of physics which much more closely resemble ecology (e.g. geophysics, hydrology, biophysics, and perhaps quantum physics), but these are not the models of science to which ecologists seem to aspire.

Third, as with our earlier analysis of the ecologists' struggle with scientific and philosophical terms and the possibility that these difficulties are not unique to ecology, we do not mean to suggest that other scientific fields are bereft of problems concerning the concepts that we will explore in this section. Complexity, variability, historicity, and other such problems may well vex some biologists, chemists, and physicists, but it is the relative importance of these issues, not their being unique to ecology, that is our concern. If, in fact, these challenges and our analyses extend to other fields (and in some cases, we've tried to cautiously make such extrapolations) such that the philosophical development we've undertaken is relevant to other sciences, then so much the better.

Fourth and finally, our comparison of ecology to physics emphasizes entities rather than processes. We are not exclusive in this regard, as relationships and processes find an accessory place in the discussion. We might have taken the tack of the process philosopher, accepting that nature's doings are primary and its entities are derivative. This might even accord well with aspects of ecology, but as alluring as this ontology might be, ecologists (and other scientists) have traditionally focused on material objects in the world – and our vocabulary, concepts, and methods are much more highly developed in this perspective. Moreover, this approach better fits the way in which ecologists have come to understand physics as a science that we should seek to emulate.

Having established the framework for our comparison of ecology and physics, the driving questions become: what aspects of ecology create, even necessitate, the conditional and particular nature of scientific claims made by ecologists; why is ecology so confusing; and is there a good reason for an inferiority complex?

The bedevilment of complexity and variability

To begin with a simple but defining consideration, ecology faces the essential problem of enormous complexity. Conceptual and empirical messiness are necessary properties of living entities and their relationships. That is, cells, organisms, species, populations, communities, and ecosystems are inherently variable, while electrons – to choose an archetypal physical entity – are fundamentally uniform. Indeed, there are at least four sources of complexity among ecological entities that need not concern the physics of subatomic particles (of course, not all planets or pendulums are the same, but the variation among particular instantiations can be explained by the concatenation of invariant physical laws).

The first level of complexity concerns the fact that living organisms exhibit genetic variation. Ecologists accept that individuals within a species are non-identical by virtue of differing genotypes, and the resulting phenotypic variation can be important with respect to ecological relations and processes. Imagine the difficulties for physics if the parentage (physical origination) of electrons yielded subatomic particles with greater and lesser masses. What if electrons from large atomic weight elements were bigger than those from hydrogen or helium? But the problem is even more difficult in philosophical terms.

The physicist who studies mechanical properties knows that actual structures vary, but can appeal to the traditional Platonic tactic of conceptualizing the ideal form. That is, while all real pendulums fail to behave like an ideal pendulum, the real-world clutter of air resistance, a rod with mass, a pivot with friction and other factors can be deemed irrelevant to the true essence of the pendulum and idealized away in order to capture the dynamics with a simple, mathematical equation. The empirical ecologist has no such philosophical escape. While the rationalist might appeal to a Platonic ideal (e.g. the population equations of Lotka and Volterra), we rarely understand what factors can be deemed irrelevant to an ecological system (chemostats and other such artificial contexts notwithstanding). As such, the ecologist cannot justifiably set aside troubling aspects of the system as part of a sweeping *ceteris paribus* clause. There is no appeal to an ideal lodgepole pine (what would be its height, its soil environment, its distance from another tree?) or snowshoe hare (would it be a male or a female, what age, what size?). The ecologist might use statistical expressions to capture aspects of an organism, but this is a very different matter from being able to assert that the empirically derived values approximate the essence of the organism.

The second level of complexity pertains to the sheer diversity of living forms and functions. The ecologist must accept the existence of millions of different kinds of entities, each with particular ecological functions and with the capacity to generate unique relationships with the others. Of course, ecologists engage in taxonomic lumping for the purposes of tractability, but this simplification is always done with some reluctance and an admission that potentially important information is lost. Studies conducted at the level of the genus or family might suggest crude patterns, but we'd be understandably hesitant to accept the findings as a reliable description or explanation of many ecological states or processes. The classical, physical scientist that captures the imagination of ecologists needs to deal with only a handful of fundamental forces and a small array of elements. Consider where physics might be today if matter was composed of 10 million different and irreducible particles, each with its own particular properties. How far along would chemistry be if the periodic table had more than 3000 rows and columns? It might look a lot more like ecology.

The third complexity is that ecologists are strapped with environmental heterogeneity. That is, even if we assumed genetically uniform populations comprised a fully resolved community of just a few dozen species, the ecological processes and relationships would vary as a function of the natural setting. After all, the outcomes of competitive exclusion experiments with just two species depend on the environmental parameters as well as initial conditions.[7] Temperature, humidity, diet, light, and other abiotic factors can substantively alter the ecology of a system – and let us not forget that within each of these types of inputs, the rate of change, duration of conditions, range, mean, etc. can all yield different outcomes (e.g. rapid cold hardening in insects[8]). Compare this complexity to the challenge of determining the properties of our exemplar electron. The physicist need not worry that a subatomic particle will alter its charge or mass as a function of its environment. Electrons associated with uranium do not differ from those found in carbon.

Box 3.2
It all depends: spatiotemporal contingency in ecosystems

Not only do keystone species change in a contingent manner as
Lawton suggested, but the relative abundances of feeding guilds are
highly dependent on ecological conditions. As shown here, spatial
and temporal changes in community structure are evident from eight
sites sampled over 9 months along 48 km of an agriculturally
impacted stream in Idaho (from Delong, M. D. and M. A. Brusven.
1998. Macroinvertebrate community structure along the longitudinal
gradient of an agriculturally impacted stream. *Journal Environmental
Management*, **22**: 445–457).

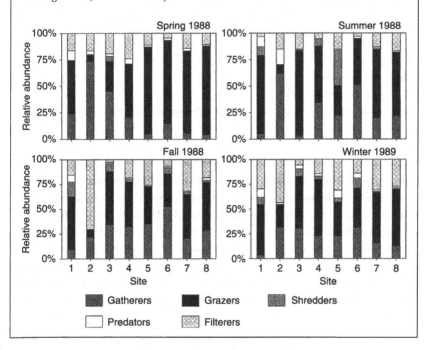

To bring the problem of context into stark relief, consider Lawton's
cogent ecological argument[9]: given the diversity of species' distributions,
we cannot expect keystone species in one community to play the same
role in another setting with different biotic and abiotic conditions
(see Box 3.2).

On the other hand, the physicist can expect electrons to do the same thing whether they are studied in Asia, Africa, or North America, or in the spring, summer, or winter. Indeed, physics operates under the assumption of universality, the axiom that physical laws and properties are invariant across the entire universe. This untested and perhaps untestable principle provides enormous simplification. Imagine the complexity if subatomic particles differed in their properties and relations as a function of their cosmic habitats.

The fourth kind of complexity that the ecologist must take into account concerns multiple response levels of the relationship, process, or system being studied. As Pickett *et al.* recognized, "ecological systems are rarely governed by one dominant cause, so the strategy for integration in ecology must deal effectively with interacting multiple causes."[10] Likewise, Allen and Hoekstra claim that in a hierarchical context, interactions of interest at level N will often be explained by mechanisms at level $N - 1$ and be constrained by processes at level $N + 1$.[11] Even this is a gross simplification, as it is evident that level $N - 1$ can constrain and level $N + 1$ can explain as well.

The responses of predators provide an exemplary case of the challenge facing ecologists. Predators can exhibit functional responses, altering their behavior in accordance with prey density. Within a single generation (and even days or hours for higher vertebrates), the predators can develop search images that bias their hunting behavior. But predator populations also can show numerical responses to prey through changes in reproduction, such that the effect is apparent across two or more generations. Should conditions persist, the predator may exhibit evolutionary responses as selection changes the nature of the system. There seems to be no obvious parallel in physics to the challenge of multiple response levels. Electrons do not respond to their environment in behavioral, reproductive, and evolutionary ways.

The vexation of historicity

The relational properties and inherent variation in living systems differentiates the study of ecology from that of physics, but there is another, equally important, distinction between these sciences. One cannot hope to understand or explain the nature of living entities and their relationships without considering the historical context.[12] As noted by Pickett *et al.*, "Ecological laws must therefore take into account that they are intended

---

Box 3.3
What if a rocket's story mattered?

Imagine a rocket moving through space according to Newtonian mechanics. A physicist might readily determine the acceleration and direction of movement based on the instantaneous force applied to the rocket. But what if the history of forces mattered? That is, what if the trajectory of the rocket depended not only on the current force applied to its current movement but also on the force that was previously applied?

If every earlier push or pull on the rocket throughout its entire journey had a lingering effect on its trajectory, the physicist would have to take into account a gust of wind at the time of take-off, a nudge from the firing of a retrorocket a week ago, and a collision with a bit of space debris yesterday. That is, "one of the first realizations that the classical philosophy of science [based on physics] was incomplete resulted from its failure to consider biological systems that had persistent effects of history."

Adapted from Pickett *et al.*'s *Ecological Understanding: The Nature of Theory, the Theory of Nature* (Academic Press, 2007).

---

to apply to evolving or historically contingent parts of the universe."[13] The events prior to the ecologist's encounter with a system matter – and often a great deal. Conversely, much of physics is conducted without reference to, or concern about, origins or preceding states (see Box 3.3).

Cosmology is the obvious exception to the ahistorical nature of physics, but no equation for the mechanical advantage of a pulley or inclined plane takes into account where the weight was in the days prior to its being released or how long the potential energy was stored. And, to return to our electron, the physicist needn't think about its history, what atom it came from, or how long it spent in association with a particular proton. The ecologist's concern with history is rooted in at least four sources, none of which pertain to physics.

First, living individuals develop, such that the effects of bodily experience (fetal, environmental, diet, climate, etc.) alter the organisms and their relations with biotic and abiotic aspects of their setting. A prototypical case of developmental effects is seen in migratory locusts, where the crowding of individuals stimulates a phase change from the solitary to the gregarious form that unfolds across developmental stages. As it matures, the insect undergoes changes in physiology (gregarious females delay sexual maturation), morphology (gregarious adults have markedly differ-ent colors and greatly extended wings), and behavioral features (gregarious nymphs aggregate and march in bands, while solitary nymphs avoid one another).[14] One can reasonably posit that subatomic physics would not be so neatly resolved if at high densities electrons altered their electrical charges, increased in mass, and attracted one another.

Second, even individuals with primitive nervous systems may alter their relations with living and nonliving components of the system as a func-tion of memory, and organisms with complex brains store memories that can elaborately modify their responses and relationships. The capacity of predators and prey to learn one another's attack and defensive tactics can shift their ecological relationships in complicated ways.[15] So, while the ecologist must take into account the previous experience of both predator and prey when setting up an experiment, the physicist need not worry that an electron has figured out through repeated encounters how to avoid being readily captured by a proton or that the proton has learned how to stalk an unwary electron.

Third, ecologists have to acknowledge that the properties and principles of their field have fundamentally changed with time. There were periods in earth's history when there were no electron transport chains, no photo-autotrophs, no predators, no nitrogen fixation, etc. Not many years ago, ecologists debated the ecological responses to large-scale use of chlori-nated hydrocarbons, and today there are animated discussions concerning what happens if the planet continues to warm at a rate of $0.2\,°C$ per decade. The ecologist must come to terms with the capacity of organisms (and consequently their ecological relationships) to evolve through muta-tion, selection, drift, recombination, and gene flow. While the cosmologist recognizes that new atomic and astronomical entities have come into existence over the course of time, the essential natures of these objects and their interactions are not subject to change. Physicists assume that photons and electrons a few billion years ago had the same properties as those today (recent discussion concerning the possibility that the speed of light has not been constant throughout cosmic history has set off a firestorm,[16] which only goes to substantiate the importance of physical entities and properties having temporal constancy). What if electrons underwent natural selection such that free electrons gave rise to offspring that were increasingly adept at avoiding capture by atomic nuclei?

Fourth, the ecological world is littered with irregularities imposed by past events. Hurricane blowdowns, avalanches, insect outbreaks, epidemics, and even asteroid impacts all affect the present relationships among living organisms and between these organisms and the nonliving components of the environment. The creation of barriers to migration and gene flow are the most obvious past events that shape current ecological conditions, but even highly localized events (e.g. where and when a moose decomposed in a meadow or a beaver constructed a dam on a stream) can greatly alter our interpretation of existing patterns of microbes, plants, or animals. In standard physics, where a particle (or a projectile or a planet) has been in the past has no bearing on understanding its current, instantaneous properties (see Box 3.3). Although physical space has plenty of local discontinuities and asymmetries, we do not need to know whether an electron has passed through any such irregularities to fully characterize the particle (interesting issues arise in quantum physics with nonlocality concepts – instantaneous connectedness between apparently separate entities[17]). Even in cosmology, the most historical of the physical sciences, physicists reconstruct the story of the universe while assuming universality – that the laws of physics are invariant in space and time.

The curse (and blessing) of scale

While inherent variation and historicity serve to differentiate many of the life sciences from the physical sciences, there is a further consideration that functions to challenge ecology's attempt to make sense of the natural world in the same way as physics does. The ecologist seeks to understand biotic relations across a staggering range of spatiotemporal scales that are fundamentally irreducible to one another. While physics provides quintessential examples of systems that can be explained as mere aggregates of component elements that are invariant under modification, ecological systems are almost necessarily emergent in their properties as defined by Wimsatt (see Box 3.4).[18] That is, the output of an ecological system is changed by intersubstitution (e.g. replacing needle-and-thread grass with western wheatgrass on North American steppes), size scaling (e.g. a $1 \, m^2$ patch of forest versus $100 \, ha$), decomposition and reaggregation (e.g. the sequence in which organisms are introduced in the process of habitat restoration can be vitally important), and cooperation and inhibition (e.g. mutualistic and antagonistic relationships abound within ecological communities).

Although emergence might not present a difficulty for physics, one might contend that the unfathomably small scale of subatomic particles

---

Box 3.4
Philosophy to the rescue:
taking the mystery out of emergence

William C. Wimsatt provided one of the most clear and testable procedures for "finding emergence." A system property is emergent (not merely aggregative) to the extent – emergence being a continuous, rather than discrete feature – that it exhibits one or more of the following properties:[21]

1. *Intersubstitution*: the system property changes after rearranging the parts or interchanging any number of parts with a corresponding number of parts from a relevant equivalence class of parts.
2. *Qualitative similarity (size scaling)*: the system property changes (identity or, if it is a quantitative property, other than only in value) after adding or subtracting parts.
3. *Decomposition* and *reaggregation*: the system property changes under operations involving the disassembly and reassembly of parts.
4. *Linearity*: the system property depends on cooperative or inhibitory interactions among the parts.

---

and forces is problematical. The claims of theoretical physicists seem utterly exotic and prone to skepticism. Indeed, Giere makes this argument, but then advocates that philosophers accept the reality of biological entities and processes when it comes to the scales of genetics or neuroscience.[19] He suggests that one could remain dubious as to the objective existence of various theoretical entities, but it is not clear exactly what these might be.

We'd propose that as one ascends from the exotic scale of theoretical physics through the familiar scale of biology (one might plausibly contend that there is an objective reality to the existence of cells, tissues, and organisms – at least for pedagogical purposes) one next encounters the scale of ecology with its populations, species, habitats, communities, and ecosystems. And these large-scale ecological entities and their associated processes seem prone to the sort of skepticism that Giere associates with theoretical physics. Has anyone ever, actually *seen* a quark, a lepton, a species, or a biome?

Levin argued that scale is "the fundamental conceptual problem in ecology, if not all of science."[20] The latter part of his contention might be questioned by physicists. In fact, for systems that are fully reducible (i.e. what Wimsatt took as being "aggregative"), scale is not a compelling issue.[21] The conservation laws of physics are prime examples of aggregative, scale invariance. However, ecology labors under an embarrassment of scale-dependent riches. This challenge is keenly appreciated by ecologists, and many fine scientists have devoted significant portions of their careers to the problems that arise from this mixed blessing.

An example of ecology's plentitude of scales is embodied in a framework known as the metabolic theory of ecology (MTE) which originated in a description of how organismal traits vary with organism size and now has possible extensions to ecological systems.[22] Because MTE captures a mathematical regularity with little in the way of explanatory mechanisms that might account for the observed pattern, some have questioned whether this is a genuine theory or a phenomenological model.[23] Setting aside this issue, it is evident that this framework and precedent concepts take advantage of general rules for change in size over many orders of magnitude of spatial and functional variables. Indeed, MTE is sometimes claimed to be scale-free. But in fact, it is limited to organisms as we know them on Earth today – tentative applications to ecological entities notwithstanding. More importantly, direct measurements of individual organisms can depart from predictions based on this theory by at least an order of magnitude.[24] As such, the MTE has a certain degree of rough utility for some sorts of problems of concern to ecologists. However, its resolution for individual cases may be so low that applications to localized problems may be fruitless.

Despite the laudable efforts of those pursuing MTE and other such scale-insensitive frameworks, no general solution to creating theories that cross ranges of spatial, temporal, or organizational scales has been found. Both Pickett *et al.* and Mitchell explicitly recognize that ecological theories have disjunct spatiotemporal domains.[25] As such, they advocate a non-mathematical integration of concepts, although exactly how ecologists might bridge across qualitatively different spatiotemporal scales and their emergent properties is not evident. As an alternative to pursuing a unification leading to an "ecological theory of everything," Borrett *et al.* have attempted to devise a system to match the appropriate ecological theory with particular, complex problems.[26]

As difficult as emergence and scaling are for theoreticians, they are no easier for empirical ecologists. Indeed, the lack of data across scales may be the greatest hindrance to developing, or at least testing, theoretical systems. For example, in an analysis of 100 papers on grasshopper ecology, Lockwood found that only 10 percent of the studies were conducted for periods of more than 2 years, in areas greater than 50 ha, or in assemblages of more than four species.[27] Thus, for a host of practical and conceptual reasons, the vast majority of field research consists of short-term, small-scale, and low diversity systems for which we lack the analytical tools to reliably extrapolate to larger and more realistic scales.

Scale might also be understood to include the network of factors that impinge on a given process. If so, then the struggle with the inadequacy of ecological theory to represent any particular instance of the natural world might stem from an exaggerated sense of the explanatory power of physical laws. Ecologists often fail to realize that all laws have limits, boundaries within which they are stable (i.e. apply) (see Box 3.5).[28]

We must understand that regularities such as that described by the ideal pendulum law are radically decontextualized. Like all such physical principles, the pendulum law is rife with *ceteris paribus* clauses (i.e. factors that are abstracted away by an appeal to "all other things being the same"). Physicists recognize that the rod has a mass, there is friction at the pivot, and air provides resistance, but they presume these are universally negligible factors. However, consider how the ideal pendulum law would have been expressed had it been developed by a physicist living on a planet in which the density of the atmosphere changed by an order of magnitude on a rapid and regular basis. In such a setting, the network of factors used to describe the period of a pendulum would surely include atmospheric resistance. Our point here is that ecologists often do not have the luxury of assuming away bothersome aspects of the world. These contingencies must be made explicit rather than quietly buried as *ceteris paribus* conditions.

Physicists, whether through humility, dissimulation, or cleverness, simply do not try to predict the course of nature at the scales typically encountered by ecologists.[29] This ecological context has been called the "middle numbers problem."[30] This problem and the solution developed by physicists was succinctly described by Pickett *et al.*: "To ecologists, the simplicity of physical laws stems from the observation that they describe either single items or statistical behaviors of many identical items."[31] That is, tractability can be attained by framing problems in terms of very few or very many elements. Systems with few entities can be fully resolved by developing equations that take into account each component and its relationship to the others. For example, a single gravitational equation incorporating the masses and velocities of the planets can be used to reliably account for the orbital patterns over cosmically short, but

---

Box 3.5
If all laws have limits, what are the
contingencies of ecological laws?

Science philosophers have long debated the nature of laws, often in
the context of classical physics. We may take "law" as equivalent to
"theory" in ecological usage. Mitchell's (2000) treatment of the
dimensions of scientific laws distinguishes between normative criteria
versus a pragmatic approach. This distinction is based upon the fact
that all laws have limits, and biological laws tend to be highly contingent
on space-time circumstances. The normative characterization of laws
is that they "allow us to explain, predict, and successfully intervene in
the world." Their features include: "(1) logical contingency (have
empirical content); (2) universality (cover all space and time); (3) truth
(exceptionless), and (4) natural necessity (not accidental)." In fact,
not all physical laws fulfill these criteria, biological laws are limited to
this planet, and ecological laws are highly contingent.

Mitchell erects a three-dimensional model for the extent to which
a law may be predictive. In this multi-factorial environment, some
laws have broader application than others. She terms the breadth
of application "stability" and states that stability is an ontological
property defining the extent along a continuum of broadening
space-time conditions over which a law or theory pertains.

We have simplified her model for stability in the following diagram
to show how a continuum of stability might apply to ecological
theories. The choices and positions of these theories are for illustrative
purposes only. We realize they could be debated at great length.

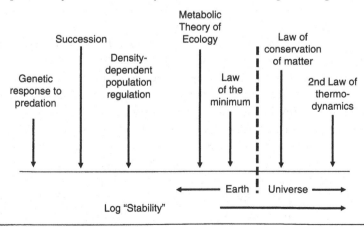

Box 3.6
Morals and middle-numbers: the challenge of ethical theory

Few philosophers have explicitly framed their work in terms of the middle-number problem. In *Nichomachean Ethics*, Aristotle admonished us to attend to the dimensions of time, place, and people in resolving moral problems. His approach of virtue ethics avoided the middle-number problem through a tactical imprecision that did not need to account for all contingencies, but avoided glossing over relevant differences. The best-known frameworks in ethics implicitly wrestle with this problem.

Converting ethics into a small-number system: Deontologists derive their claims by generalizing from one (or a few) fundamental principle(s). This rationalist approach avoids the middle-number problem, but risks oversimplification. Immanuel Kant reduced the messiness of moral action to a single notion: A person is duty bound to act with rational intent. This single moral imperative overlooked the complexity of real-life situations in which lying (Kant maintained that one could not rationally lie) yields good outcomes.

Converting ethics into a large-number system: Utilitarians avoid the middle-number problem by taking into account all individuals, at least in principle. If ethics is a matter of producing the greatest good for the greatest number, a person need not act from principle. Such an approach meant that an egregious harm to an innocent person could be ethical as long as the collective was better off. Averaging across people to smooth out local anomalies simplified ethical analyses, but overlooked concern for individual rights.

---

ecologically long, periods of time.[32] Conversely, for systems with large numbers of entities or events, the scientist relies on using the average dynamics to smooth out local irregularities. Statistical mechanics allow the physicist to treat gases as enormous collections of particles, such that the behavior of any individual molecule, no matter how aberrant, is of no concern to understanding the entire assemblage. In effect, the large number of parts are collapsed into a few average values so the system is treated as a small-number problem. The middle-number systems of ecology have too many parts to treat each one separately (as with the planets) and too few components to derive reliable averages that would subsume individuality (as with gas molecules) (see Box 3.6).

Box 3.7
**Trading meaning for precision
in understanding complex systems**

Lotfi Zadeh, the father of "fuzzy logic," recognized that multivariate heterogeneity of classes characterized many complex systems and were best dealt with as fuzzy entities. From this recognition Zadeh developed his Law of Incompatibility, illustrated below. The failure for general theory to be predictable for particular cases illustrates the impossibility for broad theories to predict locally. This is particularly true with ecological theories where local contingency rules.

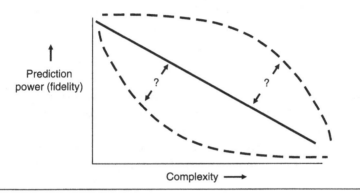

While middle-number systems appear to have particularly intractable features, Zadeh has proposed a more general scaling issue with direct implications for ecologists (see Box 3.7): "As the complexity of a system increases, our ability to make precise yet significant statements about its behavior diminishes until a threshold is reached beyond which precision and significance (or relevance) become almost mutually exclusive characteristics."[33]

From this statement, others have derived the so-called Law of Incompatibility: as complexity rises, precise statements lose meaning and meaningful statements lose precision.[34] This tradeoff is familiar to ecologists, who in the development of models find themselves compromising, rather than maximizing, among concerns for precision, realism, and generality.[35]

Of course, this concept pertains not merely to the number of entities or scale of the system but to the complexity of the relationships as well. So, a very large number of particles (e.g. a gas) would not qualify as a complex system in that the types of relationships do not increase with the quantity of molecules. Thus, many ecological systems may be in the double-bind of being both middle-number systems that preclude prediction for the reasons raised by Weinberg and complex systems whose characteristics cannot be described in terms that are both precise and relevant to our understanding for the reasons proposed by Zadeh.[36]

A perfect storm: variation, history and scale

We have made the case that ecology is different from physics in terms of the diversity of entities and processes, the importance of past events (organismal, evolutionary, and ecological), and the complexities of scale (particularly emergence and middle-number systems; see Box 3.8). Other sciences also struggle with these challenges. For example, Cleland has provided an elegant argument for why "historical science is not inferior to classical experimental science."[37] Cleland's concerns focus on the historical hypotheses associated with paleontology, archaeology, geology, planetary science, astronomy, and astrophysics (e.g. the Big Bang, continental drift, and the meteorite-impact extinction of dinosaurs), but her arguments pertain equally well to ecology insofar as explanations evoke earlier events. In addition, although Cleland dealt explicitly only with the issue

---

Box 3.8
Breaking the laws: physics in the real world

Although we claim that the conceptual approaches to physics and ecology are different in important ways, Pickett *et al.* (*Ecological Understanding*, p. 80) made the simple but clear case that even the laws of classical physics, when applied to real-world objects, are not as powerful as one might be led to believe from textbooks:

> If a committed experimentalist threw different objects out the window of a 10-storey building, she or he would find that almost none of the objects behave in conformity with the laws of gravity. Whether paper, feathers, lead balls, or a boomerang, each object would fall at different speed acceleration, and would follow a different trajectory or not fall immediately at all. This is because each object is affected by other forces such as wind, its own aerodynamic lift, and gravity in a unique combination. Contingency rules. Clearly, a law of physics shows weaknesses that are commonly believed to be typical in ecology. Perhaps the differences between physics and ecology are overrated.

Box 3.9
How experimental science really works (and fails)

As elucidated by Cleland, classical forms of science begin with a hypothesis, H (e.g. all copper expands when heated). From here one infers a testable implication, T, which states what must happen in an experiment if H is true. For example, T takes the form: if condition C (heating copper) is brought about, then event E (the expansion of copper) must occur. Using the results of an experiment in which C brought about E, the scientist has two choices:

Induction: Francis Bacon first proposed that if there is a large number of tests in which C led to E, then the scientific community should accept H. The problem with this was raised by David Hume and has yet to be solved. In short, there is no logical basis for induction; no finite amount of data is sufficient to universalize a claim.

Falsification: Karl Popper argued that the process of science involves generating stringent tests of H so that if even a single experiment fails to find that E follows from C, then the scientific community should reject H. As such, the function of science is to attempt the disconfirmation of H. There are two problems with this notion.

Auxiliary hypotheses: No experiment isolates a single hypothesis, as there is invariably a large and complex network of assumptions about the observers, equipment, and conditions (e.g. the copper didn't expand because the heating unit failed to work properly).

Actual practice: Scientists rarely practice falsification. When C does not result in E, researchers typically search for alternative explanations and conditions rather than rejecting H (e.g. checking the heating equipment, doubting the technician's competency, or questioning the actual composition of the putative copper).

---

of historicity, a careful analysis of her argument reveals that the other two problems confronting ecology (diversity and scale) are embedded in her trenchant defense of nonexperimental sciences.

Summarizing Cleland's analysis, let us begin with her most explicit contention: historicity. As opposed to the classical, experimental sciences (see Box 3.9), geology, cosmology, ecology, and related sciences attempt to explain "observable phenomena in terms of unobservable [past] causes that cannot be replicated in a laboratory setting."[38] As such, the epistemic approaches of experimental and historical sciences necessarily differ. While the former sciences focus on testing a single hypothesis under controlled conditions, the latter sciences develop a set of plausible hypotheses and then rigorously search for strongly differentiating evidence that would disconfirm all but one of these accounts (or uniquely confirm one of the hypotheses).

For example, the extinction of the dinosaurs was variously hypothesized to have been the consequence of disease, climate change, volcanism, and meteor impact. When high concentrations of iridium and shocked quartz were found at the K-T boundary – the time at which dinosaurs were found to have gone into a precipitous decline – the evidence was both supportive of the meteor-impact hypothesis and wholly inconsistent with the alternative explanations.

One might further contend that our capacity to make justified inferences about the past and to form beliefs from contemporary evidence precisely accords with a criminal investigation and subsequent trial. If a society can legitimately send people to prison – or to their deaths – based on a backward inference to the best explanation, then surely we can form credible scientific beliefs in a parallel fashion, albeit often with both a greater number of contingencies and fewer moral repercussions. Indeed, it is worth noting that Cleland refers to compelling evidence that differentiates among historical hypotheses as a "smoking gun."

Turning to the matters of diversity and scale as they pertain to ecology – and, as we will see, many of the other historical sciences – consider once again Cleland's succinct description of the historical sciences. These fields of inquiry entail explaining contemporary phenomena (e.g. the fossil record showing that the dinosaurs rapidly declined or the pattern of species on the landscape) in terms of past events, "that cannot be fully replicated in a laboratory setting." Of course, this contention applies equally to field cages, outdoor plots, and other such settings common in ecological research. Although Cleland does not make this argument, we would maintain that the reasons why scientists can't experimentally reproduce past events often pertain to diversity and scale.

The diversity of entities and processes involved in some earlier event may well preclude any contemporary effort to replicate the conditions and thereby experimentally test a particular hypothesis that could account for the phenomenon of interest. Even rather large-scale systems might be amenable to experimental methods if there were few sorts of entities engaged in a limited number of interactions. In effect, a computer simulation would become an experiment-like investigation under such circumstances. However, as Cleland notes, the reason why such modeling does not constitute an experimental test is because the investigator cannot include all of the potentially relevant conditions that prevailed. She notes:

> The most a computer can do is determine the consequences of a hypothesis under a small number of explicitly represented hypothetical conditions. It cannot determine which of these hypothetical conditions actually exists in the concrete physical system being modeled, nor can it represent all the other, possibly relevant, physical conditions present in the concrete physical system.[39]

In general, the variation among entities and processes increases with scale, so reproducing historical, large-scale phenomena is usually impossible. There is simply no way to conduct an experiment on the universe (cosmology), the earth (planetary science), a continent (geology), or the Greater Yellowstone Ecosystem (ecology). At these scales there is no way to replicate a treatment, let alone find a control (pre-treatment conditions can be taken as something of an experimental control, but this time-series approach is fraught with problems). Even if we don't add more sorts of entities, just increasing the number of biological agents in a system can lead to the emergence of new processes (e.g. the interactions among ten trees, ants, or humans are qualitatively different than those among a thousand such individuals). And finally, as either scale or diversity increases, the number of auxiliary hypotheses that a scientist must consider in interpreting the results of an experiment becomes staggering, thereby diminishing the capacity of a given investigation to (dis)confirm a hypothesis (see Box 3.9).

The implications of ecology's demons

The apparent constraints on ecological explanations do not prevent inexperienced or idealistic ecologists from aping the success of physics in finding (apparently) universal principles that explain relationships without regard to spatiotemporal context. After all, if Newton's laws were the basis for explaining both the fall of an apple and the orbit of a planet, then it would seem that a science of ecology should be able derive mathematical models for relationships which happen among microbes on the surface of an apple and among creatures moving across a planet. However, we have argued that ecology has fundamental qualities that differentiate this science from the study of nonliving systems. Moreover, important aspects of ecology distinguish it from investigations of biological systems that are highly constrained with respect to spatiotemporal scale (e.g. molecular biology, cellular biology, and physiology have all managed to bound the scale of investigation to an enviably narrow slice of time and space). In light of these considerations, ecologists must come to terms with the essential consequences.

How do we bound the scale, or extent, of our investigations? The majority of ecological studies concern localized issues,[40] while most ecological theories are generalized and global.[41] As such, the particularities of an individual case make it unlikely that overarching theories will provide specific insight to a local investigation or that the findings of a case-study will richly inform general theory. It may just be the case that the elements of variation and history preclude the development or discovery of methods to translate general theory into local terms with the specificity necessary for those attempting to resolve problems, construct explanations, or make predictions of particular ecological systems. The desire for a direct, unambiguous linkage of general ecological theory to particular ecological cases is understandable and perhaps it is worth continuing to pursue. However, believing that the inability to integrate across scales of time, space, and complexity means that a science is a conceptual weakling seems misguided. The intense, if professionally localized, interest in a philosophy of ecology suggests that we are moving away from the question of how to physicalize ecology by developing universal principles applicable to particular instances. As John Dewey argued a century ago, "intellectual progress usually occurs through sheer abandonment of questions ... We do not solve them; we get over them."[42]

As with ecology, philosophical investigations may not have the degree of exactitude that is possible with physics. Although the discipline of analytic philosophy, with its culmination in the practice of "exact philosophy," seeks to precisely define terms and convert arguments into the framework of formal logic, such an approach may not be viable for all philosophical problems. Notwithstanding deontologists' efforts to transform ethics into purely rational terms and the utilitarians' attempts to develop a "moral calculus" to optimize the good (see Box 3.6), ethical problem solving may represent a kind of middle-number system with much in common with ecology. Moral and ecological understanding does not appear to be reducible to formulae. Indeed, Aristotle admonished:[43]

> We must be content, then, in speaking of such subjects and with such premises to indicate the truth roughly and in outline, and in speaking about things which are only for the most part true and with premises of the same kind to reach conclusions that are no better. In the same spirit, therefore, should each type of statement be received; for it is the mark of an educated man to look for

precision in each class of things just so far as the nature of the subject admits; it is evidently equally foolish to accept probable reasoning from a mathematician and to demand from a rhetorician scientific proofs.

In other words, it is absurd to expect from ethics the same sort of exactitude that one seeks from physics. And likewise, a scientist who grasps the nature of of ecology with all of its contingencies and complexities will neither demand the aping of physics nor settle for sloppy rhetoric. A similar argument was made in the nineteenth century by William James with specific reference to his own field of science, psychology.[44] Just as ecology is currently castigated for its imprecision, some of James' contemporaries contended that psychology was not science. Although he fully recognized that psychology was an immature venture that had yet to attain its potential rigor and clarity as a natural science, James also held that psychology was not – and would never be – a science in the same sense as physics. And in many ways (e.g. the number of interacting elements, the richness of interconnections, the importance of evolutionary context, the role of memory, and the relevance of local contingencies), an ecosystem and its functions may be far more similar to the brain and mind than it is to a collection of gas particles.

Another approach for dealing with irreducible scales may require recognition that we are dealing with a set of irresolvable, even incommensurable, methods and theories.[45] That is, a deep-seated pluralism may be necessary in ecology. Mitchell and Dietrich have advocated exactly this approach in evolutionary biology, saying, "Rather than see a new unifying theory of evolution, however, current evolutionary biologists seem to accept multiple causal processes and types of explanations offered for evolutionary phenomena at the molecular and morphological levels. The result is explanatory and methodological pluralism in contemporary evolutionary biology."[46]

For ecology to move decisively toward understanding, predicting, or resolving important problems in the world could well require entirely different methods and techniques as a function of scale. Conceptualizing ecology in this manner will require both the constructive skepticism that has pervaded science and an element of intellectual tolerance for difference that has not always typified ecological discourse.[47] Rather than an either-or approach to theory choice (either density dependence or independence, either deterministic succession or stochastic assembly, etc.), ecologists may need to think in terms of "both-and" with the driving question being, "Under what conditions does theory A versus theory B apply?" This suggests that ecologists ought to consider abandoning the

---

Box 3.10
The really real realist: straw man or nobel laureate?

A hazard in developing any philosophical position is the creation of a straw man – a fallacious misrepresentation of an opponent's argument through the rhetorical tactics of oversimplification or exaggeration.

In this regard, one might wonder whether our stark portrayal of objectivism is an extreme caricature of a view that is not actually advocated by scholars. Giere addressed this concern and offered both a fellow philosopher's succinct definition of objectivism and a complementary excerpt from the writings of Nobel Prize laureates in physics (*Scientific Perspectivism*, 2006):

> [For the objectivist/realist], science aims to give us, in its theories, a literally true story of what the world is like; and acceptance of a scientific theory involves the belief that it is true. (Bas van Fraasen)

> We believe the world is knowable; that there are simple rules governing the behavior of matter and the evolution of the universe. We affirm that there are eternal, objective, extra-historical, socially neutral, external and universal truths. The assemblage of these truths is what we call Science ... (Stephen Weinberg and Sheldon Glashow)

---

notion that their science corresponds with a single, purely objective way that the natural world exists and functions (see Box 3.10).

Ecology's dilemma would seem to push us away from realism toward nominalism. That is, it seems incumbent upon us to adopt a philosophy in which there are no generalizations, just particular cases. As such, advancement in a complex subject area wracked by confounding influences of scale, complexity, and history forces us toward a nominalist viewpoint. Although the starting point for ecological inquiry may be the particular instance, problem, or context, there must be some intermediate ground that avoids the slippery slope toward the extreme. We endorse an approach that resembles the cautious theology of a doubter who acknowledges a mystery that can be partially grasped but never fully explained – rather than either a creed for the singular meaning of existence or a doctrine of unmitigated nihilism.

Possibly if ecologists keep foremost the conclusions of Mitchell (all laws have limits), or of Zadeh (generality and predictiveness are inversely related; Box 3.7), or of Levins[48] (generality and precision are opposing virtues), and then proceed from there with humility, it would be a good start. A pragmatic approach might be to first identify the domain in space and time (scale) for which the causal relationship of interest pertains, then constrain the appropriate theory to the extent necessary for that case, and finally use the results of the study to link the central theory to related theories so as to create useful conceptual clusters that are fruitful, testable, coherent, parsimonious, and productive in terms of the original question. As Pickett *et al.* have said: "Contingencies rule" – but they need not become dictators. This seems often to be the approach that applied ecologists, in fact, take to develop understanding of and prescriptions for specific problems in the context of a broader understanding of systems with relevant similarities.[49] We might even consider the possibility that the difference between basic and applied ecological research is not "who is paying the bill" or whether the result is economically important, but where the space-time contingency limits are located on Zadeh's curve (Box 3.7). As such, it might be more useful to see basic and applied ecological research as a continuum of situational distinction rather than somehow ideologically different.

But what makes ecology a science rather than an art? If the ecologist's perspective is so central, ecological inquiry seems to risk collapsing into human subjectivity. Is the work of manuscript reviewers and grant panels tantamount to art criticism? If there is no "right" way to view reality, what would prevent ecology from lapsing into an "anything goes" sort of venture?[50]

We shall address this vital question in greater depth toward the end of this book, but for now we propose to keep the wolves of relativism at bay through unapologetic pragmatism. Empirical adequacy devoid of the claim to knowing the one way the world "*really*" is" may be entirely sufficient to hold the center of ecology.

An appeal to pragmatism, a philosophical system in which what matters is whether our understanding leads to the satisfaction or frustration of human needs and desires, is an admission that ecological claims are anthropogenic (although they need not be anthropocentric, if we decide that other species are of material, moral, or aesthetic value). As such, we'd invert Giere's analysis of how scientists, or at least ecologists, approach the problem of scale. Using maps as an exemplar case, he suggests that: "Once the constraints of scale are accommodated, the primary reason for including some features and not others is, in most cases, the *purpose* for which the map is being made. What are the intended uses of this map?"[51]

Giere recognizes that maps are not objectively correct, but can only be judged with regard to the interests of the intended users. We'd wholeheartedly agree with this position, but suggest that Giere's account of the relationship between scale and purpose should be inverted. That is, within ecology (and perhaps other sciences) the first step of the scientific process is one of committing to a particular purpose, either implicitly or explicitly, and this becomes the constraint that must be accommodated in selecting the most appropriate scale for the interests of the researcher. Indeed, much of the philosophical confusion in ecology may be related to an attempt to deny – and thereby condemn ecologists to being constitutionally unaware of – the important role of subjectivity. Consider the following way in which this human-nature interaction can be manifest.

To begin, Keller and Golley recognize that ecology is infused with both political and aesthetic influences, such that one could argue that the science of ecology would not exist, or at least enjoy its current social and cultural status, without the influences of environmentalism and romanticism.[52] Next, the direct personal experience of ecologists with their subject matter, even a bodily immersion in the objects of study, mitigates against objectivity. We suspect that few ecologists choose this field without having a proclivity, even a deep sympathy, for the condition of the natural world. Thus, the political and aesthetic aspects of ecology infuse the individual as much as the history of the science.

Even if a scientist chose ecology in a manner utterly devoid of personal interest or passion, the type of ecologist that one becomes dramatically alters the perception of nature. Walking through Yellowstone National Park, a microbial ecologist, stream ecologist, fire ecologist, population ecologist, insect ecologist, community ecologist, vegetation ecologist, landscape ecologist, and paleoecologist would all see unique aspects of the natural world, ask distinct questions, use very dissimilar methods, and even assume that different entities and processes were real. The language and terms that these ecologists use would be underpinned by powerful conceptual structures that shape their experience. The inescapable metaphors that they use to think and speak would both entail and obscure aspects of the world.[53] Consider the implications of framing our experience of biotic relationships with such evocative terms as: communities (as if organisms lived in harmonious neighborhoods), guilds (as if groups of species had certain professions for which they trained), niches (as if there were small, concave recesses in the world that held species), exotics (as if organisms were strikingly or dramatically foreign) and invasions (as if military tactics had brought a species to a locale).

Finally, as we emphasized at the outset of this section, ecology is a science premised on the conceptual primacy of relationships. Perhaps for some scientists, Descartes' *cogito ergo sum* (I think, therefore I am) might be sufficient in so far as an entity – at least a conscious being – can bootstrap itself into existence and access certain knowledge independent of its interactions with the world. However, for ecology, the more appropriate phrase might be *afficit ergo est* (it affects, therefore it is). To exist, for the ecologist, is to have an effect. Indeed, the same may turn out to be true for physicists in that the Copenhagen interpretation of quantum mechanics suggests that a subatomic particle doesn't have any actual location or velocity until it is measured (observed) – a relationship which collapses the wave function and actualizes the particle's potential properties.

The possibility of collecting data in an objective, nonrelational manner seems "scientific," but conducting the entirety of ecology in this manner has an absurd quality, as if the ecologist's goal is to have no effect on, and to be unaffected by, that which is studied. But, in that much of ecology is driven by a concern for, or at least an interest in, the natural world, such a systemic nonrelationality seems implausible. To the extent that scientists aspire to gain understanding, insight, influence, or satisfaction from their studies, subjective experience and interest are unavoidable. So, rather than aspiring to an ecological or relational nonexistence, the ecologist might better become consciously aware of the nature of his or her relationship with that which is studied and the ways in which this relationship affects the questions posed, the methods used, and the interpretations made.

Ultimately, much of the confusion about the descriptive nature and prescriptive goals of ecology appears to be grounded in a nascent philosophy of our science. This is not to say that ecologists are philosophically naïve or unsophisticated. Rather, oftentimes the concepts necessary to construct a solid foundation for ecology are either: (1) adopted from the philosophy of physics where different issues obtain or (2) adapted in inconsistent and inexact ways among those struggling to understand what ecology is and ought to be.

Given the depth and complexity of the challenges facing ecology, the authors of this book desire to become part of the solution, rather than merely adding to the problem. But to contribute to a conceptually credible solution (i.e. an understanding of ecology that is both descriptively plausible and prescriptively compelling to fellow scientists), we must first provide a brief philosophical framework. Once we have clarified concepts and terms in this overview of philosophy, we will attempt to derive an approach that we believe takes ecology along a constructive path.

Endnotes

1. Kuhn, T. S. 1970. *The Structure of Scientific Revolutions*. Chicago: University of Chicago Press.
2. Martinez del Rio, M. 2008. Metabolic theory or metabolic models? *Trends in Ecology and Evolution*, **23**: 256–260, 256.
3. To be more accurate, our comparison primarily pertains to what might be called middle-scale physics, ranging from mechanics (e.g. levers and inclined planes) to particle physics (atoms and electrons). At finer spatial resolutions (e.g. the subatomic scale of quantum physics) individuated entities and deterministic processes give way to the still-mysterious observer effect that was famously illustrated by Schrödinger's cat. At larger resolutions, we see the emergence of perspective vis-à-vis relativity (high speeds) and the difficulties of enormously complex and unreplicable phenomena. As such, much of the literature concerning geophysics, astrophysics and atmospheric physics sounds more like ecology than it does classical, middle-scale physics.
4. James, W. 1892. A plea for psychology as a 'natural science'. *Philosophical Review*, **1**: 146–153.
5. Lockwood, D. R. 2007. Ecology is not rocket science. *Complexity and Organization*, **9**: 107–119; Pickett, S. T. A., J. Kolasa and C. G. Jones. 2007. *Ecological Understanding: The Nature of Theory, the Theory of Nature*. 2nd edn. San Diego, CA: Academic Press.
6. Gasper, P. 1991. Causation and explanation. In *The Philosophy of Science*, eds. R. Boyd, P. Gasper, and J. D. Torut. Cambridge, MA: MIT Press, 545.
7. Allee, W. C., A. E. Emerson, O. T. Park and K. P. Schmidt. 1949. *Principles of Animal Ecology*. Philadelphia: W. B. Saunders Co.
8. Qi, X. and L. Kang. 2003. Rapid cold hardening process of insects and its ecologically adaptive significance. *Progress in Natural Science*, **13**: 641–647.
9. Lawton, J. H. 1999. Are there general laws in ecology? *Oikos*, **84**: 177–192.
10. Pickett *et al.*, *Ecological Understanding*, p. XX.
11. Allen, T. F. H. and T. W. Hoekstra. 1992. *Toward a Unified Ecology*. New York: Columbia University Press.
12. Thompson J. N., O. J. Reichman, P. J. Morin, *et al.*, 2001. Frontiers of ecology. *BioScience*, **51**: 15–25; Mayr, E. 1982. *The Growth of Biological Thought*. Cambridge, MA: The Belknap Press of Harvard University Press.
13. Pickett *et al.*, *Ecological Understanding*, 21.
14. Uvarov, B. 1977. *Grasshoppers and Locusts: A Handbook of General Acridology*. 2. London: Centre for Overseas Pest Research.
15. Stanford, C. B. and R. Wrangham. 1998. *Chimpanzee and Red Colobus: The Ecology of Predator and Prey*. Cambridge, MA: Harvard University Press.
16. Webb, J. 2003. Are the laws of nature changing with time? *Physics World*, April: 33–38.
17. Grib, A. A. and W. A. Rodriguez. 1999. *Nonlocality in Quantum Physics*. New York: Springer.
18. Wimsatt, W. C. 1997. Aggregativity: reductive heuristics for finding emergence. *Philosophy of Science*, **64**: 372–384.
19. Giere, R. N. 2006. *Scientific Perspectivism*. Chicago: University of Chicago Press.
20. Levin, S. A. 1992. The problem of pattern and scale in ecology. *Ecology*, **73**: 1944.
21. Wimsatt, Aggregativity.
22. Gillooly, J. F. and A. P. Allen. 2007. Changes in body temperature influence the scaling of VO_{2max} and aerobic scope in mammals. *Biological Letters*,

3: 99–102; West, G. B. and J. H. Brown. 2004. Life's universal scaling laws. *Physics Today*, **57**: 36–42.
23. Martinez del Rio, Metabolic theory or metabolic models?
24. Ibid.
25. Pickett *et al.*, *Ecological Understanding*; Mitchell, S. D. 2000. Dimensions of scientific law. *Philosophy of Science*, **67**: 242–265.
26. Borrett, S. R., W. Bridewell, P. Langley and K. R. Arrigo. 2007. A method for representing and developing process models. *Ecological Complexity*, **4**: 1–12.
27. Lockwood, J. A. 1997. Rangeland grasshopper ecology. In *Bionomics of Grasshoppers, Katydids and Their Kin*, eds. S. K. Gangwere, M. C. Muralirangan and M. Muralirangan. London: CAB International, 83–10.
28. Mitchell, Dimensions of scientific law.
29. Lawton, Are there general laws in ecology?
30. Weinberg, G. M. 1975. *An Introduction to General Systems Thinking*. New York: Wiley; Allen, T. F. H and T. B. Starr. 1982. *Hierarchy Perspectives for Ecological Complexity*. Chicago: University of Chicago Press.
31. Pickett *et al.*, *Ecological Understanding*, 80.
32. Allen and Hoekstra, *Toward a Unified Ecology*.
33. Zadeh, L. A. 1973. Outline of a new approach to the analysis of complex systems and decision processes. *IEEE Transactions on Systems, Man, and Cybernetics*, SMC-3: 28–44.
34. McNeill, D. and P. Freiberger. 1993. *Fuzzy Logic*. New York: Simon and Schuster.
35. This concept was advanced by Levins, R. 1966. The strategy of model building in population biology. *American Scientist*, **54**: 421–431. A rebuttal to Levin's argument was given by Orzack, S. and E. Sober. 1993. A critical assessment of Levins' "The Strategy of Model Building". *Quarterly Review of Biology*, **68**: 534–546. And a defense of Levin's notion of tradeoffs, based on his concern for pragmatic features of model building not merely the formal properties of models, was given by Odenbaugh, J. 2003. Complex systems, trade-offs, and theoretical population biology: Richard Levin's "Strategy of Model Building in Population Biology" revisited. *Philosophy of Science*, **70**: 1496–1507.
36. Allen and Starr, *Hierarchy Perspectives for Ecological Complexity*; Zadeh, Outline of a new approach to the analysis of complex systems and decision processes.
37. Cleland, C. E. 2001. Historical science, experimental science, and the scientific method. *Geology*, **29**: 987–990, 987.
38. Ibid.
39. Ibid, 989.
40. Pickett *et al.*, *Ecological Understanding*; Schrader-Frechette, K. S. and E. D. McCoy, eds. 1993. *Method in Ecology. Strategies for Conservation*. Cambridge, UK: Cambridge University Press.
41. Lawton, Are there general laws in ecology?; Brown J. H. and B. A. Maurer. 1989. Macroecology: the division of food and space among species on continents. *Science*, **243**: 1145–1150.
42. Hickman, L. A. 2007. *The Influence of Darwin on Philosophy and Other Essays in Contemporary Thought*. Carbondale, IL: Southern Illinois University Press, 11.
43. *Nichomachean Ethics*, Book I, Section iii from http://classics.mit.edu/ Aristotle/ nicomachaen.html (accessed July 1, 2007).

44. James, A plea for psychology as a 'natural science'.
45. As noted earlier, we take incommensurability to describe the situation in which two approaches do not produce the same results.
46. Mitchell, S. D. and M. R. Dietrich. 2006. Integration without unification: an argument for pluralism in the biological science. *The American Naturalist*, **168**(supplement): 73–79, 76.
47. For example, see Schoener, T. W. 1986. Overview: kinds of ecological communities – ecology becomes pluralistic. In *Community Ecology*, eds. J. Diamond and T. J. Case. New York: Harper and Row, 467–479.
48. Levins, R. 1968. *Evolution in Changing Environments: Some Theoretical Explorations*. Princeton, NJ: Princeton University Press.
49. Schrader-Frechette and McCoy, *Method in Ecology*.
50. Feyerabend, P. 1993. *Against Method*. New York: Verso.
51. Giere, *Scientific Perspectivism*, 76.
52. Keller, D. R. and F. B. Golley. 2000. *The Philosophy of Ecology: From Science to Synthesis*. Athens, GA: University of Georgia Press.
53. Lakoff, G. and M. Johnson. 1980. *Metaphors We Live By*. Chicago: University of Chicago Press.

4

Finding ourselves in philosophical terms

Ecologist, heal thyself

Having explored the origins of ecology's conceptual confusion, we have achieved the equivalent of a diagnosis. We have a reasonable understanding of the symptoms and know something about the malady that afflicts our science. But, as with the practice of medicine, it is incumbent upon us to gain the participation of the patient in the course of treatment. It is not sufficient to simply declare that ecology swallow the philosophical pill that we prescribe, for the health of the field is more a matter of changing perspectives and behavior rather than merely taking an abstract elixir. So, we need ecologists to understand how their existing and robust concepts have been formalized by philosophers and how we intend to use these terms so that our proposed treatment can be adopted – or rejected – based on the scholarly equivalent of informed consent. Most readers will already understand some of the concepts that we introduce in this chapter, many will have at least passing familiarity with all of these realms, and a few will have a firm grasp of all the ideas that we present. For those with a strong background in philosophy, this overview is useful in that philosophers do not fully agree on how all of these concepts should be framed and there is no consensus on precise definitions. In addition, it will be useful to both see how we frame the science of ecology in philosophical terms and understand our use of particular concepts.

A walk in the philosophical park

Imagine an ecologist stopping alongside the road to watch a wolf approach a herd of bison in the Lamar Valley of Yellowstone National

---

Box 4.1
A stalk in the park

This image captures the essential qualities of the scenario that we use in our philosophical primer for ecologists (photo courtesy of Ranjit Warrier from flickr.com/photos/mosilager).

Park (Box 4.1). How might this scenario be captured in terms of its philosophical context, particularly as her experience pertains to the science of ecology? One way of unpacking this encounter with the living world is through dissecting the event with regard to the major realms of philosophy: beauty, right, and truth (Box 4.2).

The meaning of beauty

Attempts to answer the question, "What is beautiful?" fall into the philosophical field of aesthetics (Box 4.3). We often think of this realm as pertaining to works of art, and perhaps our ecologist takes out her camera, frames an image that appeals to her sense of balance, proportionality, color, and shading, and then takes a picture of the world. Indeed, how many field ecologists have not pursued the making of art, albeit justified in terms of an anticipated presentation to students or colleagues?

Box 4.2
A framework for philosophy

This is one of the possible ways of organizing the major realms of philosophy. Another – and perhaps somewhat more traditional – approach would be to parse the field into axiology (ethical, aesthetic, and religious values), epistemology, and ontology/metaphysics.
We prefer to give greater weight to aesthetics, given the importance of beauty to the broad enterprise of ecology as described by Keller and Golley (2000. *Philosophy of Ecology: From Science to Synthesis.* Athens: University of Georgia Press).

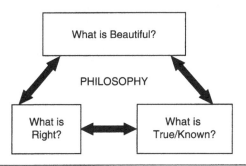

Also, explicitly present in our scenario is the natural world, which possesses its own aesthetic qualities. The beauty of nature is a topic of intense interest among aestheticians, and there is perhaps no better source to begin an exploration of this important field than Allen Carlson's compelling work.[1] Ecologists will likely be pleased to find that Carlson argues for the importance of having a scientific understanding of the natural world to claim genuine aesthetic appreciation.

Finally, let us allow that as the ecologist watches the bison respond to the presence of the wolf, she forms in her mind a new insight regarding the long-standing question concerning predator energetics as well as a set of experiments that will disprove her hypothesis. Indeed, her planned experiments draw together principles in a way that nobody has previously conceived. Would we be surprised if she described her idea as an "elegant solution," and might we even assert that her approach suggests a "beautiful, new theory" of predator ecology? Indeed, scientists have an abiding sense of aesthetics with respect to ideas, experiencing visceral pleasure

Box 4.3
A framework for philosophy and aesthetics

The realm of aesthetics includes the analysis of beauty with respect to art, nature, and ideas. Whether there is a single conception that unites these three contexts is not entirely clear. Further pursuit of this concept is out of place here, although a sense of pleasure, well-being, or satisfaction seems to arise in the presence of a beautiful painting, sunset, and equation – at least for those who understand how these reflect on our humanity. We propose that most ecologists see beauty in nature and that perception has, or has had, a role to play in how they have come to be in ecology, the subjects they pursue, and the value they place on their science.

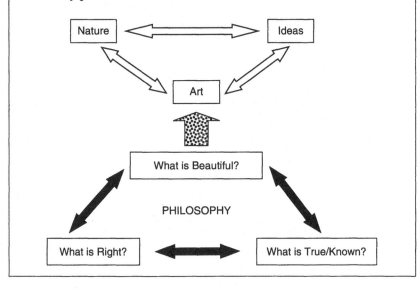

in concepts that accord with qualities strikingly similar to artworks: simplicity, symmetry, balance, refinement, unification, richness, harmony, etc. Although ecologists possess and contribute to an aesthetic appreciation of human artifacts, nature, and ideas, our contemplative scientist probably does not consider herself primarily engaged in asking, "What is beautiful?" But then, she may be only slightly more aware of the extent to which the second, major thread of philosophy – ethics – is woven into her science.

The meaning of right

To ask "What is right?" is to engage in another branch of philosophy (Box 4.4). A common starting point for addressing this question can be termed "axiology" (the study of values) and scientists are wholly enmeshed in this sort of discourse. Our ecologist is likely mulling over many possible lines of investigation. For each of these, she is likely engaged in assessing if a particular question is worth her time, energy, and talents; if a hypothesis is important enough to warrant funding; and if a set of experiments is of enough value to justify a dissertation-level study for her graduate student. All of these are axiological issues, but some lines of inquiry – the questions concerning moral value – are particularly compelling in that they set the stage for ethical decisions.

Being in Yellowstone, the ecologist might reasonably wonder whether the reintroduction of wolves was the right thing to do. She could frame this analysis in terms of whether the program yielded a favorable outcome for the organisms in, and visitors to, the Park. From this consequentialist perspective she might embrace a utilitarian ethic in which the right action is that which provides the greatest good for the greatest number. On the other hand, she might hold that humans, who extirpated the wolves, had a moral duty to restore these animals to the region (or that the wolves had a right to live there). From such a deontological perspective, she may well believe that our intentions are at issue. What matters is whether we rationally determined what principle was relevant and acted out of this duty; if so, we are ethical regardless of the results. Or perhaps she maintains that humans should live virtuously, such that the reintroduction of the wolves was an act reflecting the properties of courage, temperance, and fairness. If we live in accordance with these properties, then ethical actions will naturally flow for the virtue ethicist. The point being that once the ecologist asks what we *ought* to do, she is engaged in a moral question.

Now that the wolves are in the Park, our ecologist might raise questions of justice which extend to the collective. In particular, she might ask how we should go about fairly distributing the costs and benefits of the predators. Should ranchers be expected to absorb the costs of predation on livestock? Who should pay for monitoring the wolf population? How should the Park allocate access to the best viewing areas? Perhaps the ecologist's concerns for justice pertain to the distribution of scientific resources, leading her to wonder whether she will be fairly treated when she proposes her grand new concept to a funding panel – will the

Box 4.4
A framework for philosophy and the right

The matter of how we ought to conduct our lives can be addressed
through the disciplines of ethics, justice, and political philosophy.
These aspects of living a good life pertain to different scales: the
individual, the society, and the state – and clearly a reductionistic
approach fails. Scholars of justice have noted that we cannot assure
that doing the right thing at the individual level will assure right
actions by society. Collectives have different concerns, abilities,
and obligations than do individuals (e.g. while no individual has
a duty to be a doctor, teacher, or police officer, a just society is
obligated to provide for the health, education, and protection of
its citizens).

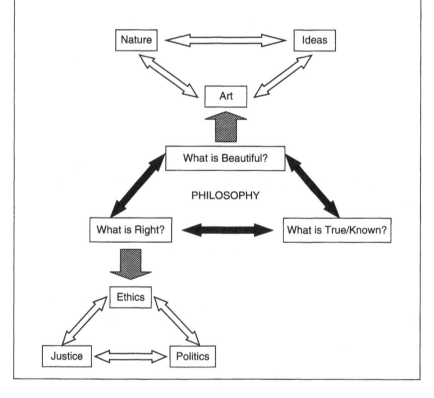

limited research dollars be distributed with respect to the principle of need (favoring scientists from smaller, poorly supported institutions), merit (favoring scientists with the best ideas and records of productivity), ability (favoring scientists most likely to deliver research products), compensation (favoring scientists who have been unfairly excluded from support in the past), or some other considerations? In short, the foundation for just distribution of resources is a matter of keen interest to all researchers.

From here, our ecologist might question whether a federal agency, such as the National Science Foundation, should even be engaged in the support of ecological research, whether the Department of Interior should maintain a system of National Parks, or whether the US Fish and Wildlife Service should be involved in the management of wildlife. That is she may well ask, "What is the proper role of government?" Her political philosophy may favor authoritarianism based on her belief that people tend to act badly, or perhaps she is highly suspicious of government and advocates libertarianism. In any case, it is quite impossible for our ecologist to avoid matters of political philosophy. She cannot avoid entangling science and government if she receives public funding, accesses public lands, educates students at a public institution, or finds her work enabled or restricted by laws and policies.

The meaning of truth

As much as our ecologist is entangled in the first two grand questions of philosophy, she probably sees matters of beauty and right as complicating the central question that drives her science. She would likely say that the pursuit of the truth is her reason for being a scientist. And the question, "What is true?" will be the focus of the balance of our book and the foundation for the philosophy of ecology that we will propose. But embedded in this vital question are three philosophical inquiries that must be understood and addressed (Box 4.5).

First, we must ask the question of ontology: "What exists?" That is, our ecologist must begin with an ontological commitment, an affirmation of what entities and processes in the Lamar Valley exist independent of her experience. She might assert that blades of grass and individual bison are really there, but she may be doubtful of species, niches, communities, and ecosystems. If a nominalist, she might contend that all abstract groupings are human inventions, that there is no way to be sure that such categories conform to actuality. As an essentialist (or realist) she'd argue that we

Box 4.5
A framework for philosophy and truth

Perhaps the most conceptually challenging questions in philosophy pertain to ascertaining what is real, what properties or qualities reality possesses, and how we know the answers to these questions. However, these pivotal philosophical issues do not often find their way into ecological discourse, textbooks, journals, or conferences. Rather, the nature of reality is often taken to be "given" – assumed to be shared among ecologists as a kind of collective axiom. But we've found that when pressed, ecologists evince a much more diverse set of positions than might be presumed.

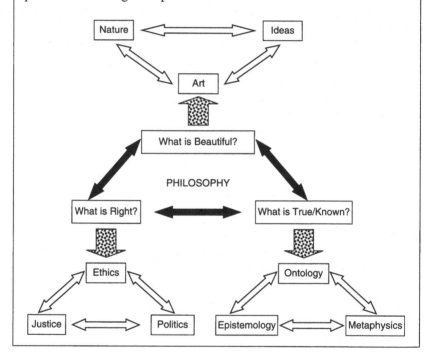

can "carve nature at its joints," so that natural kinds such as herds, populations, and meadows, and their properties correspond to mind-independent realities.

She might also be dubious of the ontological primacy of the material world and hold that processes are fundamental to existence. Objects or entities may be understood as the observable evidence of underlying processes, such that a wave on the surface of a lake is merely the physical manifestation of a flow of energy or a bison is the biological process of breathing, eating, metabolizing, mating, and otherwise functioning as that which we call "bison." Whether she considers entities or processes as ontologically primary, the existence of both (and their differences in kind) might appear to our ecologist as being essential.

If ontology is like the taxonomy of reality, then metaphysics – which asks, "What is the nature of that which exists?" – might be thought of as the anatomy, physiology, behavior, and ecology of existence. The nature of reality, its fundamental properties and relations, demand the attention of the scientist. For the philosopher, the question of whether the world is composed of one substance (monism) or different kinds of entities (e.g. mind and body, as in a long-standing version of dualism) is a central concern, although other issues may be of greater interest to our ecologist. She might, for example, hold that there are pure forms of existence of which the actual world is but a messy representation (e.g. idealized populations accord with logistic growth, although actual populations, such as Yellowstone's bison, are buffeted by complicating, extraneous factors). From this perspective, things and the properties that they possess out of biophysical necessity are of metaphysical concern. Or, she could see the natural world as analogous to the river that carved the Lamar Valley, believing that there is nothing but ceaseless change. If so, she might further assert that process or relationality (e.g. the interaction of the wolf and bison) is what underwrites the science of ecology. Another view would be that we can't know anything about reality in itself because our perceptions of the world are just that – subjective mental experiences that may not accord with inaccessible, external existence. If so, our dubious ecologist may eschew any knowledge of objective reality. No matter her metaphysical position, she must take the third step in the pursuit of truth.

Epistemology answers the question, "How do we know?" – exploring the nature, source, limits, and justifications of what we take to be true beliefs. It is likely that our ecologist would claim to know about the world through both rationalism (truths derived from axioms, such as the idea that populations cannot grow indefinitely) and empiricism (truths via sensory experience, such as her knowledge that bison are herbivores). If the ecologist further asserts that what she knows accords with objective reality, then she ascribes to a correspondence theory of truth as do many of her colleagues. As a scientific realist, she accepts that the referents of scientific terms (e.g. individuals, populations, and species) exist in a mind-independent world.

But is our ecologist betraying science if she doubts whether such a claim can really be sustained under harsh scrutiny – that she actually knows about this reality outside of her own perceptions – and ascribes to a version of anti-realism? Not at all. She might be comfortable with the claim that scientific knowledge is empirically adequate, that mid-scale observable entities correspond with reality, but the theoretical constructs or unobservable entities that concern much of ecology may not correspond with the world (e.g. populations, communities, and ecosystems are unobservable insofar as their boundaries are vague and arbitrarily delimited, and their individual elements are spatiotemporally transitory and cannot be simultaneously observed in the collective). However, scientists invariably apply the standard of coherence to claims of knowledge. That is, the grand concept of predator energetics that she has been contemplating is exciting both because it is a new perspective and because it fits in with, and even unifies, aspects of ecological science. But as with many of her colleagues, our ecologist might maintain that at least some of her claims correspond to reality, thereby assuring that science is more than an internally coherent fiction.

It is possible, albeit rather unlikely, that our ecologist would entirely abandon realism. Her doubt may have begun with seeing the fundamental problem with induction, the form of inference on which science rests but which lacks an epistemic, rational justification. Or perhaps she is simply unconvinced that science is anything more than a social construction, a cultural fabrication with no connection to objective reality – if such even exists. This post-modernist approach is the contemporary version of skepticism, rejecting the idea that we can have knowledge of anything other than our own mental states. But solipsism seems to offer little to the ecologist who wishes to take seriously the work of science. Although some would assert that the scientist must choose between a hard-driving absolutism and a nihilistic relativism, there is a rich, philosophical middle ground between these extremes.[2]

Our ecologist might maintain that there is a real world that gives rise to, and constrains, our subjective experiences. But this staggeringly rich and complex existence can only be known through our relationship to it. As such, she could hold that truth is a matter of meeting human needs and desires. As a pragmatist, she might take the position that various sciences constitute different domains of knowledge, within which each can make legitimate claims of truth insofar as the practitioners are concerned. In other words, the veracity of her "beautiful new theory" will be affirmed if it meets the desires of scientists for simple, fruitful, consistent, and comprehensive explanations, provides wildlife managers with useful predictions upon which to base policies and practices that sustain valued organisms, or allows the public to access a part of nature in a way that accords with psychological, cultural, and economic interests.

This analysis of philosophy has taken us to the upshot of this book. Can a formulation of pragmatism – a kind of mitigated realism – provide a plausible descriptive and prescriptive foundation for the ontological, metaphysical, and epistemological issues of ecology?

Endnotes

1. Carlson, A. 2000. *Aesthetics and the Environment: The Appreciation of Nature, Art, and Architecture*. New York: Routledge.
2. Giere, R. N. 2006. *Scientific Perspectivism*. Chicago: University of Chicago Press.

5

Ecological pragmatism and constrained perspectivism: an introduction

Many perspectives, but not any perspective

To address the conceptual and philosophical issues within ecology, we propose that one possible, coherent approach is a form of ecological pragmatism that might best be termed, *constrained perspectivism*. In brief, perspectivism holds that when different observers have different interests (needs, desires, and capacities which may be conscious or not) the individuals may make claims about the world that are irreducible and unresolvable. Moreover, acting on these claims can yield incommensurable outcomes that fail to satisfy the interests of all individuals. A <u>constrained</u> perspectivism has the important caveat that the different claims can represent partial truths reflecting different properties of reality that objectively limit the empirical verifiability of the propositions. This view recently has been expressed by Wimsatt who argued:[1] "[T]his multiple rootedness need not lead to 'anything goes' perspectival relativism, or an anti-naturalist worship of common sense, experience, or language. It yields a kind of multi-perspectival realism anchored in the heterogeneity of 'piecewise' complementary approaches common in biology and the study of complex systems."

Being rooted in the philosophy of pragmatism, a framework of constrained perspectivism further posits that we are justified in holding that a claim is true if acting upon the belief produces consequences that satisfy genuine desires (i.e. those that actually accord with our well-being, such that an addict's craving for a drug is a proximate desire but is not within that individual's ultimate interests; Box 5.1). As such, we needn't worry that science is trivialized – indeed, quite the opposite according to William James:[2]

> Let me begin by reminding you of the fact that the possession of true thought meant everywhere the possession of invaluable instruments of action; and

that our duty to gain truth, so far from being a blank command from out of the blue, or a "stunt" self-imposed by our intellect, can account for itself by excellent practical reasons ... The possession of truth, so far from being here an end to itself, is only a preliminary means toward other vital satisfactions.

Box 5.1
Satisfying our desires

Aristotle recognized both epistemic (mental) and physical (bodily) cravings, the satisfaction of which might be taken to represent the goals of basic and applied science. The American pragmatist John Dewey saw this as an unhelpful dualism between mind and body. It has been shown recently that knowing produces a sense of physical pleasure (see latimes.com/news/opinion/la-oe-biederman19-2008 jul19,0,3327488.story). Conversely, we might also contend that satisfaction of bodily desire often entails knowing (except in such matters as rolling over in our sleep). The most explicit connection between scientific knowledge and the satisfaction of desire was made by William James who argued ("What pragmatism means," in *Pragmatism, Old and New*, pp. 297, 319.):

> ... *ideas (which themselves are but parts of our experience) become true just in so far as they help us to get into satisfactory relation with other parts of our experience*, to summarize them and get about among them by conceptual short-cuts instead of following the interminable succession of particular phenomena. Any idea upon which we can ride, so to speak; an idea that will carry us prosperously from any one part of our experience to any other part, linking things satisfactorily, working securely, simplifying, saving labor; it is true for just so much, true in so far forth, true *instrumentally*.

> Dewey, Schiller and their allies, in reaching this general conception of all truth, have only followed the example of geologists, biologists, and philologists. In the establishment of these other sciences, the successful stroke was always to take some simple process actually observable in operation – such as denudation by weather, say, or variation from parental type, or change of dialect by incorporation of new words and pronunciations – and then to generalize it, making it apply to all times, and produce great results by summating its effects through the ages.

> ... Truth in science is what gives us the maximum possible of satisfactions, taste included, but consistency both with previous truth and with novel fact is always the most imperious claimant.

Giere's scientific perspectivism

Our philosophy of ecology is largely consistent with Ronald Giere's recent work on scientific perspectivism,[3] but there are some substantive divergencess from – or perhaps conceptual extensions of – his highly compelling analysis. Given the importance of Giere's work and its general compatibility with our contentions, it is worth considering the central similarities and differences. Like Giere, we hope to mediate between the "strong objectivism of most scientists, or the hard realism of many philosophers of science, and the constructivism found largely among historians and sociologists of science."[4]

Central to Giere's project is the proposition that scientific claims are conditional along these lines: "According to this highly confirmed theory (or reliable instrument), the world seems to be roughly such and such."[5] If so, there is no way to take the further step of asserting a complete or literally correct picture of the world. While this seems to be reasonable, we are seeking both more and less of ecology. We want more in that at least some claims are not merely rough approximations. As scientists, our assertions are stringently constrained by objective reality so that while within these limitations there may be a roughness (although precision seems possible as well) there also is a kind of precision imposed by the way the world actually is. But we also want less than Giere's scientific perspectivism offers, insofar as ecologists often lack highly confirmed theories. So, our claims about parts of the world are not always cast in terms of some law-like principles and may even be highly local, idio-syncratic, and disconnected from an overarching theory. This worldview of ecologists is in tension with that of physical absolutism, a difference that John Dewey explicated:[6]

> It is an old remark that human progress is a zigzag affair. The idea of a universal reign of law, based on properties immutably inhering in things and of the nature as to be capable of exact mathematical statement was a sublime idea ... From this point of view, the principle of indeterminacy seems like an intellectual catastrophe. In compelling surrender of the doctrine of exact and immutable laws describing the fixed antecedent of things, it seems to involve abandonment of the idea that the world is fundamentally intelligible. A universe in which fixed laws do not make possible exact predictions seems from the older standpoint to be a world in which disorder reigns.

Giere notes that previous work by philosophers points to the conclusion that reality can sustain more than one account of it and even potentially

---

Box 5.2
Formulations of pluralism

Imagine that there appears to be a penny resting on its edge on a table, such that one observer (A) sees the "heads" side of the coin and another observer (B) sees "tails". The following represent possible versions of pluralism.

Intra-observer pluralism: Observer A moves around the table until the coin appears as a vertical line. So, through a change in time or space a single individual can perceive the same object in two different ways.

Epistemic pluralism or *Perspectivism*: Observer A describes the coin as bearing Lincoln's silhouette, while observer B describes the coin as bearing the Lincoln Memorial. Differences in these accounts might be resolved via communication between the observers.

Sensory pluralism: Observer A describes the coin as copper-colored. Observer B, who suffers from a form of color-blindness, describes the coin as grey.

Cognitive pluralism: Observer A, who remembers Lincoln's assassination, describes the same view as a source of sadness. Observer B, who has moved next to Observer A, recalls Lincoln's courage and describes the coin as a source of hope.

Ontological pluralism: Observers A and B contend that there is no single thing on the table, agreeing that on the table is a heads-coin and a tails-coin.

Skeptical pluralism: Observer A contends that there is no actual coin as the experience is a purely subjective state of mind, while observer B insists that there is an external world with a coin.

Indeterminate pluralism: Observer A refuses to commit to "heads", maintaining that we can never really know whether our experiences correspond to things in the world. Observer B claims to know that there is a tails-coin.

---

incommensurable accounts (Box 5.2). However, he resists radical pluralism, at least for all of science. Giere argues that such extreme contingency – often reflecting social circumstances – is seen in "weak science," which he takes to be those fields that are at the limits of experimental capabilities (e.g. gravity waves). If the standard of scientific strength is experimentalism, then much of ecology would have to be considered weak. We would object to the notion that observational and model-based understanding is necessarily weaker than experimental approaches, given that the study of macroecological change, ecosystem processes, and biospheric dynamics would all be, of methodological necessity, weak sciences. The attribution of weakness to the study of entities and processes that exist at spatiotemporal scales that preclude experimentalism, seems an unfounded judgment of scientific inquiry.

Ecology's broader perspective

Giere may be correct in maintaining that some investigations are particularly prone to cultural influences, and we'd agree that an account of ecology must include the socioeconomic perspective of its practitioners (Box 5.3). As such, Giere's more limited analysis of scientific perspectivism, which is based on the biophysical properties of a person (or instrument), provides an important, but incomplete, account of ecology. In that he seems to hold that all entities in science are perceived by vision or instruments, he excludes at least two important features of ecological investigation.

First, this approach overlooks the importance of processes, rather than objects, in understanding the world. We'll explore this limitation subsequently in further detail, but for the moment it is sufficient to appreciate that one does not see scramble competition, population growth, or nutrient cycling per se, but one does observe the physical instantiations of these processes and infers the underlying dynamics.

Second, many ecological entities are not perceived (i.e. seen by our eyes or instruments), but conceived. That is to say, many of the driving questions in ecology pertain not only to the qualities of an entity (as with Giere's fascinating examples of the interactive or relational property that we call "color") but to the existence of the entities themselves. Perhaps ecology is particularly prone to ontological and metaphysical problems such that we are concerned with how to carve up the world into entities and processes that are often unobservable (has anyone actually seen species, speciation, communities, metabolism, ecosystems, or equilibrium?).

The notion that truth claims are always relative to a perspective is, as Giere points out, not particularly radical. Even the Logical Empiricists could not escape the fact that scientific claims were invariably relative

---

Box 5.3
Ecology, science and God

The importance of religion to the historical development of the environmental sciences is richly analyzed in Peter Bowler's, *The Earth Encompassed*. The particular ways in which theology has shaped the course of ecology is recounted by Elspeth Whitney in "Christianity and changing concepts of nature: an historical perspective" (2006. In *Religion and the New Ecology*. University of Notre Dame Press, pp. 26–52). In this same book, Mark Stoll's chapter, "Creating ecology: Protestants and the moral community of creation" reveals the ways in which the religious backgrounds of key ecologists in the twentieth century profoundly influenced their science.

Others have also described – and objected to – these associations. Philosopher Mark Sagoff has contended that ecology's focus on stability and equilibrium, "blurs the line between science and religion" (1997. Muddle or muddle through? Takings jurisprudence meets the Endangered Species Act. *William and Mary Law Review*, **38**: 888).

In Paul Feyerabend's *Against Method*, he develops an intriguing perspective on religion and science. Feyerabend notes that God is powerful and must be obeyed because it is a force outside of all traditions. Likewise, he argues, objectivity is tradition-independent and plays a central role in rationalism which is a "secularized form of the belief in the power of the word of God." If so, then science's appeal to the irresistible and utterly compelling force of rationality and objectivity is conceptually and historically rooted in theology.

Whatever one's view on the proper relationship between science and religion, there can be no doubt that ecology and theology are historically, conceptually, and culturally linked in ways that have profoundly shaped ecologists' axiology, ontology, and metaphysics.

---

to language. What Giere proposes is to extend this doctrine of linguistic contextualization to one's physical position (both in relation to the object of study and in terms of the internal processing of the incoming signal). He focuses on the physical characteristics of generalized models, primarily those of physics, but with some intriguing extensions into evolution and economics. While this is surely a valuable enterprise, it is too limited for constructing a viable philosophy of ecology.

We have two concerns with respect to the scope of Giere's analysis. First, ecology has few generalized models, although Giere's analysis would certainly be interesting to apply to the metabolic theory of ecology (MTE)[7] or Lotka-Volterra models.[8,9] As such, a philosophy of ecology will need to account for those elements of the science that are local and not embedded in generalized models. Second, while ecologists certainly have physical positions in the world from which they make observations (or use instruments to do so), this is not a sufficient account of perspectivism within our science. While Giere conditionalizes statements with, "From where we stand" in a literal sense (our physical position gives rise to a particular visual perspective), we extend "From where we stand" to its metaphorical sense. That is, we stand in relation to certain interests (personal, social, cultural, etc.) and values (the importance of future generations, the interests of beauty, and the centrality of economics). It is this kind of standing that must also be taken into account by ecologists.

Finally, Giere notes that in physics, models are constructed according to explicitly formulated principles. He then plausibly contends that even some life sciences and social sciences have such principles; evolution has natural selection and economics has equilibrium. To the pragmatists, the reason why scientists of all stripes are compelled to seek such constancy is rooted in a deeply seated human concern that was perhaps most evident in classical physics, but continues to pervade our thinking:[10]

> It would be hard to find a franker statement of the motive which controlled Newton's doctrine. There was some guarantee that Nature would not go to pieces and be dissipated or revert to chaos. How could the unity of anything be secure unless there was something persistent and unchanging behind all change? Without such fixed indissoluble unities, no final certainty was possible. Everything was put in peril of dissolution. These metaphysical fears rather than any experimental evidence determined the nature of the fundamental assumptions of Newton [and] furnished the premises which he regarded as scientific and as the very foundation of the possibility of science.

These principles are sometimes called empirical laws, generalizations that are universal and true. But, Giere points out, understood in this way, these assertions are either vacuously true or known to be false. While he sees this as a problem, we are rather less concerned for two reasons.

First, all of ecology is "false" if we take this to mean that to be true is provide the whole truth and nothing but the truth. Of course, our claims about the nature of the world are partial, but there is an important difference between being false and being incompletely correct. To say that grasshoppers are herbivores is largely correct, but it is false insofar as these insects are also necrophagic scavengers. Moreover, the pragmatist would have little patience with those who insist that statements correspond to the way the world "really is." For even if we were to accept such a notion, we'd have no way of knowing if our claims have this sort of one-to-one match with objective reality. According to constrained pespectivism, at best we can know if our claims fail to accord with some limit of the external world. As such, we adopt a Popperian-like stance in which reality can falsify our claims, but it cannot affirm them.[11]

Second, our concern is not with the sort of truthfulness that seems to concern Giere with regard to scientific laws or generalizations. Although he renounces scientific claims that assert a singular correspondence to an objective way that the world really is, he does appear to hold that two observers in the same place, at the same time can make true claims by virtue of intersubjective objectivity (i.e. precisely shared biophysical perspectives). However, we'd assert that one must also take into consideration experiences, interests, affinities, and psycho-social contexts – not only physics and neurology – if one hopes to have a viable philosophy of ecology. Of even greater importance, in terms of pragmatism, the falseness of a claim is assured insofar as we've invariably chosen a partial account of the world. As George E. P. Box so sagely noted, "All models are wrong, but some are useful."[12] And it is usefulness that allows a coherent and reasonably complete account of the science of ecology.

Some explication of what pragmatists mean by "truth" seems in order to assuage concerns that this notion has no place in constrained perspectivism. Pragmatists do not ascribe to a unified version of the truth, but there is widespread agreement as to qualities that truths possess. To begin, *truth satisfies our desires*. That is to say, "A new opinion counts as 'true' just in proportion as it gratifies the individual's desire to assimilate the novel in his experience to his belief in stock" (see also Box 5.1).[13] Given our desire for consistency, it follows that *truth is coherent*. James contended that the "greatest enemy of any one of the truths may be the rest of our truths"[14]. In other words, ideas are true, "in so far as they help us get into satisfactory relation with other parts of our experience."[15] But do truths pertain to anything in the world? *Truth has mitigated correspondence*, not a 1-to-1 match with the external world, but a correspondence with a participatory reality. Along similar lines, Charles Peirce argued that, "The opinion which is fated to be ultimately agreed to by all who investigate, is what we mean by the truth."[16] Convergence through investigation means that: *truths are ideas that survive testing*. James maintained that, "Truth *happens* to an idea. It *becomes* true, is *made* true by events,"[17] and F. C. S. Schiller simply stated that, "a truth which will not ... submit to verification, is not a truth at all."[18] From this point emerges John Dewey's contention: *truth is warranted assertability.* If we have justifiable reasons for claiming something, then we have come into relation with a truth. But what warrants our assertions? *Truth is what works.* Richard Rorty's contention that truth is the compliment we pay to ideas that work is a modern echo of James' notion that, "You can say [of an idea] that 'it is useful because it is true' or that 'it is true because it is useful.' Both these phrases mean exactly the same thing."[19] Because what is useful varies, *truth is contextual.* As Schiller argued, " 'success,' therefore, in validating a 'truth,' is a relative term, relative to the purpose with which the truth was claimed. The 'same' predication may be 'true' for me and 'false' for you, if our purposes are different."[20] Applying our ideas in particular domains serves as a counter-balance to the pragmatists' other notion that: *truths are expedient ways of thinking.* There is both utility and hazard in generalization: "[Truth must be] expedient in the long run and on the whole, of course; for what meets expediently all the experiences in sight won't necessarily meet all further experiences equally satisfactorily ... making us correct our present formulas."[21]

In sum, contrary to the exalted position that truth holds for absolutists and rationalists, for pragmatists such as Nelson Goodman, "Truth, far from being a solemn and severe master, is a docile and obedient servant. The scientist who supposes that he is single-mindedly dedicated to the search for truth deceives himself."[22]

We do not claim that our framework of constrained perspectivism is authoritative or flawless, as our purpose is not to dictate how various concepts must be used, but to clarify how they might be framed so as to provide valuable tools for ecologists. Nor do we assert that our solution is original; indeed, our proposed path relies on adapting the extensive work of previous guides. And finally, because we find that many of the conflicts in ecology have arisen and continue to persist through the construction of false dilemmas (e.g. relativism versus absolutism and realism versus nominalism), we have no interest in asserting that ours is "the" solution to the philosophical challenges of ecology, only that it is "a" solution to many important problems.

Perspectivism is not relativism

A few words concerning relativism are in order to clarify constrained perspectivism. The paradigmatic case of relativism for ecologists appears to be the work of Joan Roughgarden.[23] Robert McIntosh has pointed out the inconsistencies of Roughgarden's arguments, such as her castigating critics of competition theory for holding untenable philosophical positions, and then also arguing that being concerned with a philosophical position during scientific research can lead to bad science.[24] The notion that we ought to keep our philosophy implicit, rather than explicit, does seem to be a dangerous intellectual strategy, but Roughgarden's point was not so much that we ought to be studiously oblivious to our philosophical position, but that ecologists ought not to systematically exclude alternative views by virtue of our philosophical commitments.

This distinction becomes clearer as Robert Peters draws the link between Roughgarden's contention and the philosophy of Paul Feyerabend:[25] "Roughgarden *et al.* (1989) suggest the anarchism of Feyerabend (1975) is an appropriate guide for pluralistic ecology. Feyerabend is a radical because he does not separate synthesis from analysis ... from this, it follows that anything goes."

Peters has adopted a common, but badly mistaken, interpretation of the complex arguments made by Feyerabend.[26,27] In fact, the "anything goes" assertion is a *reductio ad absurdum* – a "principle" that rationalists derive from Feyerabend's work if they are committed to universal principles (not unlike Dewey's critique of our psychological and cultural anxieties for an unchanging truth in *The Quest for Certainty*).

Elisabeth Lloyd has presented a much more sophisticated analysis of Feyerabend's purported relativism concerning the claims of science.[28] She argues that Feyerabend promoted vigorous testing of scientific claims against all other possibilities (in this context, "anything goes"). The purpose of this continuous challenging was to prevent science from lapsing into a self-satisfied dogma with a socially sanctioned position of assumed power.

So when Peters asserts that, "For the working scientist, the richest sources of inspiration are the traditions of science, personal scientific experience and the larger culture in which these are imbedded,"[29] Feyerabend would become anxious. Such a frame of reference could well exclude alternatives by imposing the dominant – and often unexamined – traditional and cultural standards. Protagoras warned that one should not presume that one's own village and customs (or an ecologist's subdiscipline or familiar theoretical context) are "the navel of the world."[30] His admonition was not an endorsement of relativism, but an affirmation of humility and perspectivism.

Pluralism, guided by human interests and constrained by the real world, seems to be a descriptively accurate account of, and a prescriptively compelling direction for, ecology.[31] An unconstrained relativism in the form of a true "anything goes" approach to truth is as untenable for pragmatists as it was for Feyerabend.[32] This may be how the world appears to a rationalist confronting the metaphysical fear that without a single, absolute, universal, timeless Truth "everything would be put in peril of dissolution."[33] However, the pragmatist is not panicked by a world resting on the backs of contingencies that go all the way down.

So we reject forms of relativism that degenerate into intellectual nihilism and thereby eschew the possibility of authentic knowledge. But neither do we promote a false syncretism that unites all of ecological thinking under a single construct. For advocates of a pragmatic pluralism (a logical outcome of constrained perspectivism) to contend that they have the one-and-only solution to the philosophical complexities of ecology would be hypocrisy and folly.

Perhaps the best way of understanding the potential of constrained perspectivism for addressing the concerns of ecologists is to systematically analyze the approach in terms of the conventional elements associated with the philosophy of science: ontology, metaphysics, and epistemology.[34] In these contexts, we will explore how constrained perspectivism draws on its pragmatic roots to come to an understanding of what exists, what properties real entities and processes possess, and how we can know about the nature of the world.

Endnotes

1. Wimsatt, W. C. 2007. *Re-Engineering Philosophy for Limited Beings: Piecewise Approximations to Reality.* Cambridge, MA: Harvard University Press.
2. James, W. 2006 [1907]. Pragmatism's conception of the truth. In *Pragmatism Old and New: Selected Writings*, ed. S. Haack. Amherst, NY: Prometheus, 312.
3. Giere, R. N. 2006. *Scientific Perspectivism.* Chicago: University of Chicago Press.
4. Ibid, 3.
5. Ibid, 6.
6. Dewey, J. 1960 [1929]. *The Quest for Certainty: A Study of the Relation of Knowledge and Action.* New York: Capricorn, 208.
7. Whitfield, J. 2006. *In the Beat of a Heart: Life, Energy, and the Unity of Nature.* New York: The National Academies Press.
8. Lotka, A. J. 1920. Undamped oscillations derived from the law of mass action. *Journal of the American Chemical Society,* **42**: 1595–1599.
9. Lotka, A. J. 1956. *Elements of Mathematical Biology.* New York: Dover.

10. Dewey, *The Quest for Certainty*, 117.
11. Popper, K. 2002 [1959]. *The Logic of Scientific Discovery*. New York: Routledge.
12. See Wired Magazine on-line, wired.com/science/discoveries/magazine/16–07/pb_theory, accessed 10 October, 2008.
13. James, W. 2006 [1906]. What pragmatism means. In *Pragmatism Old and New: Selected Writings*, ed. S. Haack. Amherst, NY: Prometheus, 299.
14. James, What pragmatism means, 306.
15. Ibid, 297.
16. Peirce, C. S. 2006 [1878]. How to make our ideas clear. In *Pragmatism Old and New: Selected Writings*, ed. S. Haack. Amherst, NY: Prometheus, 147.
17. James, Pragmatism's conception of the truth, 315.
18. Schiller, F. C. S. 1907. *Studies in Humanism*. New York: Norton, 8.
19. James, Pragmatism's conception of the truth, 313.
20. Schiller, F. C. S. 2006 [1907]. The making of truth. In *Pragmatism Old and New: Selected Writings*, ed. S. Haack. Amherst, NY: Prometheus, 499.
21. James, Pragmatism's conception of the truth, 322.
22. Goodman, N. 2006 [1975]. Words, works, worlds. In *Pragmatism Old and New: Selected Writings*, ed. S. Haack. Amherst, NY: Prometheus, 613.
23. Roughgarden, J. 1984. Competition and theory in community ecology. *American Naturalist*, **122**: 583–601.
24. McIntosh, R. P. 1987. Pluralism in ecology. *Annual Review of Ecology and Systematics*, **18**: 321–341.
25. Peters, R. H. 1991. *A Critique for Ecology*, New York: Cambridge University Press, 25.
26. Feyerabend, P. K. 1975. *Against Method: Outline of an Anarchistic Theory of Knowledge*. London: Verso.
27. Feyerabend, P. K. 1978. *Science in a Free Society*. London: Verso.
28. Lloyd, E. A. 1997. Feyerabend, Mill, and pluralism. *Philosophy of Science*, **64**: 396–407.
29. Peters, *A Critique for Ecology*, 23.
30. Feyerabend, *Science in a Free Society*, 28.
31. We do not wish to imply that we are being original in proposing a pluralistic ecology. Robert McIntosh's "Pluralism in ecology" (1987. *Annual Review of Ecology and Systematics*, **18**: 321–341) pre-dates our work by two decades. We would suggest, however, that ours is a more thoroughgoing pluralism both in terms of its scope and depth. McIntosh advocated a methodological pluralism for community ecology, and his arguments were well-conceived and persuasive. We are proposing that pluralism pertains to all of ecology, not just the study of communities, and that a pluralistic perspective is appropriate for axiological, ontological, and metaphysical matters, not just epistemological issues in ecology.
32. Fine, A. 2007. Relativism, pragmatism, and the practice of science. In *New Pragmatists*, ed. C. Misak. New York: Oxford University Press, 50–67.
33. Dewey, *The Quest for Certainty*, 117.
34. We do not wish to marginalize the importance of axiology – and, in particular, ethics – to the practice of science. We will allude to the moral implications that arise via constrained perspectivism, but for the most part either these are not radically different from current concepts or they parallel our other arguments (e.g. moral truth is pragmatic and perspectival, but not degenerately relative).

6

Ecological pragmatism and constrained perspectivism: ontology

Existence as contingence

From the position of constrained perspectivism, the answer to the ontological question of "what exists?" is a richer set of possibilities than would be the case from the more traditional approach of absolute scientific realism. Rather than an objective view of a singular reality, we suggest that ecologists recognize a world composed of a continuum of entities and processes.[1] We are certainly sensitive to the reaction that discussions of ontology and metaphysics might elicit from ecologists. Such dubiousness has hardly diminished in the century since William James wrote:[2]

> Refinement has its place in things, true enough. But a philosophy that breathes out nothing but refinement will never satisfy the empiricist temper of mind. It will seem rather like a monument to artificiality. So we find men of science preferring to turn their backs on metaphysics as on something altogether cloistered and spectral, and practical men shaking philosophy's dust off their feet and following the call of the wild.

This is why the pragmatists do not argue or worry about the "really real" or the existence of that which had no effect on our experience. Whether or not species are actually in the world or are constructs of human ingenuity is undecidable – and, what's more, there seems to be no consequences to the conduct of science whether one or the other is actually the case. Although Hillary Putnam's concern was about the putative objects of physics, his contentions apply equally well to populations, communities, and ecosystems:[3]

> Quine [a pragmatist philosopher of recent times] has urged us to accept the existence of abstract entities on the ground that these are indispensable in mathematics, and of microparticles and space time-points on the ground that these are indispensable in physics; and what better justification is there for accepting an ontology than its indispensability in our scientific practice?

The question, therefore, is not whether an entity or process exists in some absolute, objective, mind-independent way, but whether our positing its existence is of demonstrable utility to our endeavors. We shall know if our ontological commitment accords with reality by virtue of whether or not our acting upon an idea generates outcomes consistent with our interests, needs, and desires. For at least some pragmatists even the existence of a reality that "pushes back" is subject to the criterion of usefulness. This standard, rather than some ethereal intellectualizing is why these pragmatists are comfortable with talking about (and acting as if there truly is) a real world:[4]

> For the belief in the world theory of ordinary realism, in a 'real world' into which we are born, and which has existed 'independently' of us for æons before that event, and so cannot possibly have been made by us or any man, has very high pragmatic warrant. It is a theory which holds together and explains our experience, and can be acted on with very great success. It is adequate for almost all our purposes. It works so well that it cannot be denied a very high degree of truth.

This tension between objective existence and cultural constructs was elegantly resolved by William Wimsatt, who argued for a pan-realism in which social relativism is viewed as a part of the real world with its own potential for causal effects (see Box 2.2). His view is compatible with constrained perspectivism although it attends to somewhat different concerns regarding the complexity of science, including both the way the world really is and the way in which we have developed our conceptual and our social frameworks.[5]

An ontological commitment to ecological processes

Although we conventionally think of reality in terms of entities as the basis for reality (see Chapter 2), the importance of process as an onto-logical foundation is highly compatible with the precepts of pragmatism (Box 6.1).[6,7] The process philosopher maintains that the world is better understood in terms of its becoming than its being. The notion of "being as becoming" or the view that existence is the unending process of coming into the world may be traced to Aristotle, although his emphasis on things and substances might seem to vitiate such a contention.[8] The foundational concept for the modern process philosopher (and many pragmatists) is that framing reality in terms of dynamic modes, rather than fixed stabilities, is a more compelling perspective given that change (physical,

Box 6.1
Pragmatism and process

The original American pragmatists were not explicit in terms of their metaphysics, largely because they considered such ruminations to be generally unfruitful ponderings of something unknowable – and whatever position one took rarely had an evident effect on one's actions in the world. However, we can infer that these philosophers took processes at least as seriously as entities as a useful way of framing their own concepts. In *The Quest for Certainty* (New York: Capricorn, p. 137), John Dewey argued:

> In like fashion, thought, our conception and ideas, are designations of operations to be performed or already performed. Consequently their value is determined by the outcome of these operations.

If operations were at the heart of Dewey's pragmatism, practice – another expression of process – was key to Charles Peirce's assessment of what was worthwhile ("How to make our ideas clear," in *Pragmatism, Old and New*, p. 137):

> Thus, we come down to what is tangible and practical, as the root of every real distinction of thought, no matter how subtle it may be; and there is no distinction of meaning so fine as to consist in anything but a possible difference of practice.

And finally, William James did not deny that there are entities but contended that "leading" was a vital process for ideas to warrant our consideration ("Pragmatism's concept of truth," in *Pragmatism Old and New*, p. 318):

> Agreement thus turns out to be essentially an affair of leading – leading that is useful because it is into quarters that contain objects that are important. True ideas lead us into useful verbal and conceptual quarters as well as directly up to useful sensible termini ... The untrammeled flowing of the leading-process, its general freedom from clash and contradiction, passes for its indirect verification; but all roads lead to Rome, and in the end and eventually, all true processes must lead to the face of directly verifying sensible experiences *somewhere*, which somebody's ideas have copied.

organic, psychological) is pervasive and predominant.[9] In this light, many life scientists might embrace an inversion of the classical principle *operari sequitur esse* (function follows upon being) such that *esse sequitur operari* – form follows function. Some philosophers have differentiated between ontology (in which being is a state or static condition) and kinesthology (in which being is process or a dynamic unfolding). This distinction is not unlike the way ecologists think of homeostasis versus homeorhesis. Nicholas Rescher notes that from the perspective of process philosophy:[10] "[P]rocesses are basic and things derivative, because it takes a mental process (of separation) to extract 'things' from the blooming buzzing confusion of the world's physical processes. For process philosophy, what a thing *is* consists in what it *does*."

The notion that we make out of ceaseless change a world of things which accord with our interests is entirely consistent with pragmatism and points to another possible extension of Ronald Giere's work on scientific perspectivism.[11] He considers objects to be the metaphysical basis of scientific inquiry, such that their properties are the source of contention about our claims concerning the world. Given that we do not "see" processes per se, but rather we pick out objects that manifest a perpetually unfolding reality, it is not clear if or how Giere's focus on the importance of perspective for human or instrumental visualization would apply to the unseen. Indeed, whether one takes entities or processes to be real is, itself, a matter of perspectivism. The constrained perspectivism that we are advocating includes not only the physical location and internal processing of the observer relative to a thing in the world but the individual's social and psychological position with respect to a process. The ecologist does not see competition or cooperation. Behavioral ecologists might witness interactions that are agonistic, which are taken to represent instantiations of competition, but the energetic costs or reproductive benefits of these encounters are not directly observed. However, the scientist does choose a conceptual perspective from which to interpret relationships among organisms.

An ontological "middle way"

Whether the ecologist focuses on processes or entities, these range from the highly concrete (being principally mind-independent or "objectively discovered," such as planets and sound waves) to the highly abstract (being largely mind-dependent or "subjectively created," such as populations and carrying capacity). Between these poles is a broad spectrum of entities and processes that are conceptualized ("interactively made" as a result of the relationship between a subjective observer and objective existence, as can be exemplified with ecosystems and plant succession).

We might also think of entities, processes, and their properties as being contingent on our experiences and interests such that we "abstract away" those elements that we deem irrelevant. In arguing that, "Our capacity for overlooking is virtually unlimited, and what we do take in usually consists of significant fragments and clues that need massive supplementation,"[12] the contemporary pragmatic philosopher Nelson Goodman was echoing the words of William James:[13] "The human mind is essentially partial. It can be efficient at all only by picking out what to attend to, and ignoring everything else, – by narrowing its point of view. Otherwise, what little strength it has is dispersed, and it loses its way altogether. Man always wants his curiosity gratified for a particular purpose."

This philosophical move accords with the excoriation of unfruitful debates by other pragmatists such as John Dewey who found little worth arguing in terms of subjectivism and objectivism. Indeed, he argued that the subject-object dichotomy employed by philosophers should be replaced with an organism-environment concept.[14] Dewey was strongly influenced by the teaching of G. H. Perkins and the evolutionary writings of T. H. Huxley. As a result, Dewey's view posits an explicitly ecological and relational foundation for ontology:[15] "an object or event is always a special part, phase, or aspect, of an environing experienced world – a situation ... We live and act in connection with the existing environment, not in connection with isolated objects, even though a singular thing may be crucially significant in deciding how to respond to total environments."

Moreover, his criticism of pointless arguments about false, or at least unproductive, conceptual dichotomies might apply to at least some of the ontological debates that have consumed a great deal of intellectual resources in ecology – debates in which the outcome, if either side was to prevail, would have done little to alter the course of science. Perhaps James put the matter in the most compelling terms:[16]

> Whenever a dispute is serious, we ought to be able to show some practical difference that must follow from one side or the other's being right ... There can *be* no difference anywhere that doesn't *make* a difference elsewhere – no difference in abstract truth that doesn't express itself in a difference of concrete fact and in conduct consequent upon that fact, imposed on somebody, somehow, somewhere, and somewhen.

Such an understanding of "what exists" means that the pool of conceptualized entities and processes is far deeper than many have conceived in ecology (Box 6.2).[17] Some leading ecologists have been staunch advocates of approaches that explore the rich and messy middle-ground between dichotomous extremes. Levins and Lewontin argued that, "It is also possible to opt for compromise in the form of a liberal pluralism in which the questions become quantitative: how different and how similar are objects?[18]

Having probed the ontology of constrained perspectivism, we next turn our attention to metaphysics. If what is taken to be real is a matter of our interests, insofar as they accord with the limits imposed by the external world, then what does this mean for the properties of these entities and processes?

Box 6.2
The existence of bison

The existence of "bison" (as individuals, herds, species, and genus) can be expressed in terms of their degree of subjectivity/objectivity. As such, the x-axis of this figure captures the continuous nature of ontological claims as understood through constrained perspectivism. The y-axis is explained in the following chapter.

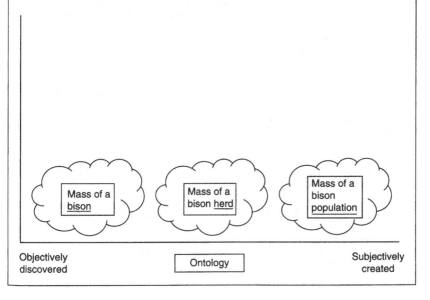

Objectively discovered   Ontology   Subjectively created

Endnotes

1. At various places in this book, we've noted the valuable contributions to the philosophy of ecology made by Steward Pickett and his colleagues (Pickett, S. T. A., J. Kolasa and C. G. Jones. 2007. *Ecological Understanding*. San Diego: Academic Press). Their work is certainly accessible to scientists, but explicit comparison to our framework is difficult because they never use the terms ontology, metaphysics, axiology, or ethics, and only once refer to epistemology. We believe that such concepts are embedded in the text, but we are reluctant to reframe their positions in philosophical terms. It appears that much of their analysis pertains to epistemology, which is the philosophical field that has traditionally been the focus of ecological critiques. Other matters, such as the nature of reality, are entangled in their epistemological approach, and we suspect that many of their views would accord with the framework of constrained perspectivism.
2. James. W. 1956 [1896]. *The Will to Believe, Human Immortality, and other Essays on Popular Philosophy*. New York: Dover, 20.
3. Putman, H. 2006 [1987]. Reality and truth. In *Pragmatism Old and New: Selected Writings*, ed. S. Haack. Amherst, NY: Prometheus, 632.
4. Schiller, F. C. S. 2006 [1907]. The making of truth. In *Pragmatism Old and New: Selected Writings*, ed. S. Haack. Amherst, NY: Prometheus, 507.
5. Wimsatt, W. C. 2007. *Re-Engineering Philosophy for Limited Beings: Piecewise Approximations to Reality*. Cambridge, MA: Harvard University Press, 148.
6. Debrock, G. 2003. *Process Pragmatism: Essays on a Quiet Philosophical Revolution*. Kenilworth, NJ: Rodopi.
7. Hausman, C. R. 1997. *Charles S. Peirce's Evolutionary Philosophy*. New York: Cambridge University Press.
8. Rescher, N. 2008. Process Philosophy, *Stanford Encyclopedia of Philosophy*, accessed 6 October 2008, plato.stanford.edu/entries/process-philosophy/
9. Rescher, N. 1996, *Process Metaphysics: An Introduction to Process Philosophy*, New York: SUNY Press.
10. Rescher, Process Philosophy, section 2.
11. Giere, R. N. 2006. *Scientific Perspectivism*. Chicago: University of Chicago Press.
12. Goodman, N. 2006 [1975]. Words, works, worlds. In *Pragmatism Old and New: Selected Writings*, ed. S. Haack. Amherst, NY: Prometheus, 61.
13. James, *Will to Believe*, 219.
14. Boisvert, R. D. 1988. *Dewey's Metaphysics*. New York: Fordham University Press.
15. Dewey, J. 2006 [1938]. Common sense and scientific inquiry. In *Pragmatism Old and New: Selected Writings*, ed. S. Haack. Amherst, NY: Prometheus, 451.
16. James, W. 2006 [1907]. What pragmatism means. In *Pragmatism Old and New: Selected Writings*, ed. S. Haack. Amherst, NY: Prometheus, 291.
17. Zellmer, A. J., T. F. H. Allen and K. Kesseboehmer. 2006. The nature of ecological complexity: a protocol for building the narrative. *Ecological Complexity*, **3**: 171–182.
18. Levins, R. and R. C. Lewontin. 1980. Dialectics and reductionism in ecology. *Synthese* **43**: 47.

7

Ecological pragmatism and constrained perspectivism: metaphysics

The properties of reality

With respect to metaphysics, constrained perspectivism holds that the properties of entities and processes are enormously diverse and vary along a continuum of mind-(in)dependence. At one end of the spectrum there are objective properties which can be discovered (e.g. our ecologist in Yellowstone might be interested in primary properties such as a bison's mass, although even mass is a relational property, being relative to the frame of the observer; Box 7.1). At the other end of the spectrum lie the attributed properties which are created by (groups of) individuals (e.g. our ecologist might perceive the nobility of *Bison*). In between these extremes are the properties which are interactively made (e.g. secondary properties such as our ecologist perceiving a bison's color or estimating the bison population's carrying capacity) (Box 7.1). Lakoff and Johnson describe interactional properties as basic to their experientialist account of truth, which lies between the objectivist and subjectivism formulations.[1] These interactional properties of an entity or process are bounded by the nature of the world, but within these bounds the nature of a property varies with the perspective of the observer.

The conditional or perspectival nature of properties that we propose is rooted in Plato's claim that to be real is to be capable of affecting and being affected.[2] That is, if something has no capacity for relationship with anything else, then it makes no sense to suggest that it is real. Insofar as human knowledge is concerned, for a property to be real means that we can experience it, so we are intractably "in the game." The pragmatists fully understood this – as well as being attuned to ecological concepts – in choosing to refer to relationships in terms of organism-environment,

---

Box 7.1
The existence of bison

The ontology of "bison" (as individuals, herds, species, and genus) is expressed on the x-axis, while the metaphysics or properties (mass, color, relatedness, nutrient requirements, wisdom, and nobility) are reflected on the y-axis. As with ontological claims, metaphysical assertions are understood to be of a continuous nature through constrained perspectivism.

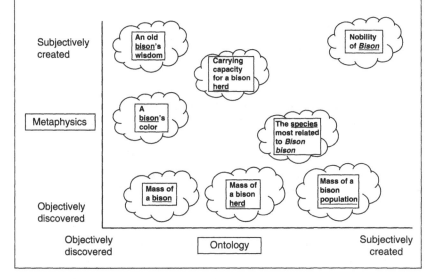

---

rather than subject-object. There is no property-in-itself; the qualities of things (and processes) exist only in context of relations.

Charles Peirce recognized that the interactions were not merely binary, but often involved an entire network of relationships. He argued persuasively that the nature of atoms depended on an entire system of observers, assumptions, mathematics, instruments, values, and histories that comprise science. This pragmatist argument was powerfully validated in 1909 via the famous double-slit experiment. The compelling empirical evidence for the interactive nature of properties (even for something as presumably simple as light) came when light was shown to have the properties of a wave or a particle depending on how the observer chose to relate to it. We are not outside spectators of nature, but participants in what we take to be the real qualities of the world.

This is not to say that pragmatists – or proponents of constrained perspectivism – give up on being objective. Rather, objectivity is seen as a regulative ideal instead of a condition that can be attained. As John Dewey admonished, "The mind must be purified *as far as humanly possible* [emphasis added] of bias and of that favoritism for one kind of conclusion rather than another which distorts observation and introduces an extraneous factor into reflection."[3] The key, therefore, is to become keenly aware of our own perspectives. Lakoff and Johnson argued that neither subjectivism nor objectivism was viable and proposed "experientialism" as a way of acknowledging that we must frame our knowledge in metaphorical terms. From within a particular system of metaphors, it is certainly possible to assess the truthfulness of a claim, but our understanding will always be domain-specific. We can rise above individual bias, but we can't ascend to a God's eye view such that truth is no longer relative to a particular conceptual system.

The profoundly interactive and relational nature of existence, properties, and knowledge is at the foundation of constrained perspectivism. This philosophy is also richly entwined with the worldview of ecologists. That is, we understand that any claim regarding the properties of entities and processes depends on where the scientist stands – both physically and in terms of his/her interests, prior experiences, and values – within the system. As Dewey contended in his account of the Copernican Revolution, which has now spread to the philosophy of ecology:[4]

> The old center was mind knowing by means of an equipment of powers complete within itself, and merely exercised upon an antecedent external material equally complete in itself. The new center is indefinite interactions taking place within a course of nature which is not fixed and complete, but which is capable of direction to new and different results through the mediation of intentional operations. Neither self nor world, neither soul nor nature (in the sense of something isolated and finished in its isolation) is the center, any more than either earth or sun is the absolute center of a single universal and necessary frame of reference. There is a moving whole of interacting parts; a center emerges wherever there is effort to change them in a particular direction.

A metaphysical "middle way"

Let us bring the discussion of metaphysics back to more tangible and biological terms. The perception of color is a classic example of the importance of interactional properties in the context of perspective.

As thoroughly explicated by Ronald Giere, colors are real, but their reality is perspectival.[5] For example, from one angle an object may appear blue, while from another it may be seen as green. Our thoughtful ecologist in the Park might not even notice that the summer pelage of bison is a lustrous brown with many spectral overtones when the sunlight comes from behind her, but the creature is a relatively flat dull color, almost a matte black, on a cloudy day. Indeed, Giere extends his argument of color vision to human perception in general and ultimately to our instruments, which are sensitive to only particular kinds of input. With respect to both the human sensory system and that of our contrivances, the "output" is a function of both the sorts of allowable inputs and the internal constitution or processing of this information.

We further point out that perspectives may also differ between species. A flower seen from the perspective of a human may be white, but it may appear as strikingly patterned from the vantage of a bee's visual system (which includes the ultraviolet wavelength or "bee purple"). As such, neither the claim of the flower being plain or striped is truer than the alternative. Likewise, the property of edibility is radically different from the perspective of a human and a tomato hornworm which can feed on various nightshades. As such, it makes no sense to argue that either the human or hornworm is wrong about the plant being poisonous (the perspectives of the two organisms yield incommensurable solutions to the problem of what to eat).

Although most of Giere's analysis concerns biophysical aspects of perspective (one's neurological apparatus and physical environment), he briefly recognizes conceptual perspectives, contending that scientists may adopt theoretical positions (e.g. Newton's laws characterize the mechanical perspective and Maxwell's laws exemplify the electromagnetic perspective). What Giere is less clear about is how one chooses or ought to go about choosing a perspective, except perhaps via accident, habit, enculturation, evolution, or other such conditions. And here is where pragmatism may provide some clarity.

Perspective depends on the observer's context, which is derived from one's interests, but these need not be conscious. Needs and wants can be evolutionarily grounded and highly compelling without our being cognizant of them. And even when desires are conscious, they may not be autonomous; our wants are not developed in a cultural vacuum. What we believe we want is often shaped by opportunities. For example, research funding can play an important role in driving the perceived desires of the scientific community.

Diversity as a metaphysical virtue of ecology

Because of a perspectival metaphysics, the ways in which the ecologists' world can be engaged are enormously varied (Box 7.1). Our ecologist standing in the Lamar Valley, is probably fascinated by the size, beauty, and power of bison. Indeed, there is good reason that we refer to these animals as charismatic megafauna. While the aesthetic properties of the animals are enticing, the ecologist's investigations may be further motivated by her ethical questions concerning how we ought to manage bison populations. That is, should they be accorded positive utilitarian value to human visitors or negative utilitarian value to livestock producers concerned about brucellosis, or simply be accorded intrinsic value? As for truth, depending on her goal, our ecologist might adopt the view that what is true about bison ecology is that which accords with human interests. Or she might prefer coherentism, and frame questions in terms of existing theories and models of wildlife, mammal, ungulate, or herbivore biology. Or our ecologist might aspire to correspondence and attempt a hard-driving objectivity with double-blind methods or even a mathematical modeling approach that is grounded in biophysical necessity rather than the fallibility of empirical sense data. But regardless of these choices, she cannot escape that her science is grounded in a particular and personal worldview.[6]

Another way of understanding the metaphysical implications of constrained perspectivism may be familiar to ecologists. We are comfortable with the notion that different organisms perceive the world in different ways such that no particular view is privileged. A female bison could be correctly classified as a competitor (by another female), a host (by a biting fly), or a prey (by a wolf). And to a male bison, she is a potential mate so even within a species the "right" view is contextual but constrained. That is, if the male perceives the female as a food item, his interests in eating will be frustrated – but even this is contingent, for if the male is a newborn and the female is his mother, she is a source of food, if not a food item. As such, it should not be surprising to find constrained perspectivism among humans having different interests, such that a bison can be validly seen as an aesthetic object (artist), a moral entity (animal rights activist), a meal (hunter), a disease vector (rancher), a herbivore (plant ecologist), a carbon sink (biogeochemist), a dung source (insect ecologist), a competitor for grass (community ecologist), or a reproductive unit (population ecologist). And so, what constitutes an interactional property is the relationship between a rich set of objective potentialities in the real world (what philosophers refer to as "dispositional properties") and an entity whose subjective interests actualize one or more of these latent possibilities. For example, it makes little sense to refer to sucrose as being sweet if there existed no organism with a sense of taste; sugar has the latent, objective potential of being sweet, but this property is only actualized through the interaction with, and experience of, another entity.

The metaphysics of constrained perspectivism may also shed some light of the nature of ecology as a science through a complementary philosophical concept – Wittgenstein's family resemblance.[7] In short, he proposed that complex subjects, such as ecology, may not be definable in terms of any necessary and sufficient properties. For example, one might try to define balls as being spherical, but this would exclude footballs, so one might refer to them as inflated objects of play but this would exclude ball bearings, and so on. But this does not mean that the concept of "ball" is vacuous. Rather, it may be the case that one ball has properties A and B, another ball has properties B and C, and a third ball has properties C and D. If so, then the first and third balls have no properties in common, but share the quality of being a ball via the properties that they have in common with the second ball. As such, balls have a family resemblance. So, what about ecology?

We'd suggest that Wittgenstein's non-absolutist approach would be a sensible way of understanding the pluralism within ecology. If ecology is a kind of family resemblance, there may be no necessary and sufficient condition for a discipline to be a kind of ecological science. Instead, population biology may have properties in common with community ecology that may, in turn, have aspects in common with ecosystem ecology that may then share features with geochemistry – but population biology and geochemistry may have no qualities in common except through the intermediate linkages with another realm of ecology. The search for a singular answer to, "what is ecology?" may be a fool's errand. Rather than "the science of ecology" we might think in terms of "the sciences of ecology."

Constraining perspectives to avoid ecological anarchy

Our ontology and metaphysics are similar in some ways to those recently proposed by Amanda Zellmer *et al.* (Box 7.2). Their contention was that science is a kind of narrative, a way of telling stories which do not need to correspond to an objective reality, nor provide internal coherence.

Box 7.2
Science as unconstrained narrative

Zellmer *et al.* (2006) showed how modeling cycles can function to create a narrative from a range of perspectives regarding salmon in the Columbia River. What, if anything, constrains the beliefs of the various stakeholders or the form of the emergent narrative is not entirely clear. As such, this way of understanding science has some important common ground with constrained perspectivism but lacks the commitment to there being a real world that empirically limits our accounts.

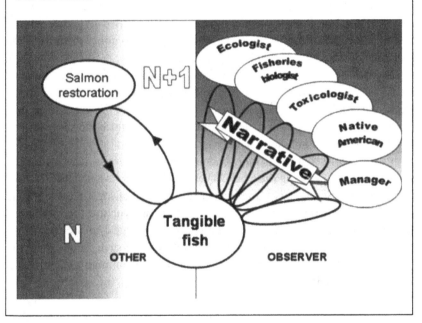

These accounts must only be useful and convincing, such that we come to believe that we are all seeing the same thing. Zellmer *et al.* contend:[8]

> A narrative [including that of science] is not about the reality of a situation. Rather, the point of the story is to lay out in the open what the narrator suggests is important. Narratives are not about being objective, but are instead displays of subjectivity ... The narrative is an expression, not of the verity of anything, but of the values that are shifting as the story unfolds.

They appear to be proposing an unconstrained perspectivism, and this strongly differentiates their view from ours or that of the pragmatists: "We give up the doctrine of objective certitude, we do not thereby give up the quest or hope of truth itself."[9]

One can see where the approach of the pragmatists allows ecologists to avoid unproductive philosophizing while being able to ground claims in something other than collective taste or emotional persuasion. Borrowing from pragmatism, constrained perspectivism imposes the standard of useful outcomes. Consider, for example, how such a philosophy deals with the problem of nominalism and realism in ecology. William James suggested that pragmatism (like ecology) has nominalistic tendencies in that it was "always appealing to particulars."[10] But he understood the utility of realism, "when we have once directly verified our ideas about one specimen of a kind, we consider ourselves free to apply them to other specimens without [complete] verification."[11]

Whether or not populations, species, and communities are "natural kinds" is not the issue. Rather, the question is whether these are "useful kinds" that allow us to act in ways that satisfy our interests. For one research project, the useful (true) taxonomic level might be species, but for another it could be families – neither is a "natural kind" if by this we mean a set or category that exists independently of our interests. The truth of a particular way of parsing the world is verified by the consequences of acting on these ideas. So, there is a nominalistic foundation of sorts in ecology and pragmatism, but this simply means that we then structure the individual entities/processes in ways that can be tested with regard to the usefulness of acting in a manner consistent with the existence of such "kinds."

So, as lyrical and appealing as Zellmer *et al.*'s vision might be, we argue that it is a sufficient description of only one end of the spectrum – the most abstract, subjectively created entities and processes. To varying degrees, the world "pushes back." Sense experience limits rational belief formation. If our Yellowstone ecologist was seeking affirmation from nature and imagined that the bison was a cuddly creature in need of human contact, she would likely find that acting on this formulation of reality would tragically fail to accord with her interests. Our stories, and scientific narratives are constrained by the world. That is to say, ecological inquiries necessarily include a significant element of – but are not wholly a matter of – subjectivism. While postmodernism's versions of cultural construction can be useful in pointing to the importance of social history in the development of ideas, to ground all explanations in such a perspective is implausible.

Endnotes

1. Lakoff, G. and J. Johnson. 1980. *Metaphors We Live By*. Chicago: University of Chicago Press.
2. Plato (N. P. White, translator). 1993. *Sophist*. Cambridge, MA: Hackett Publishing Co.
3. Dewey, J. 1960 [1929]. *The Quest for Certainty: A Study of the Relation of Knowledge and Action*. New York: Capricorn, 68.
4. Ibid, 290.
5. Giere, R. N. 2006. *Scientific Perspectivism*. Chicago: University of Chicago Press.
6. Zellmer, A. J., T. F. H. Allen and K. Kesseboehmer. 2006. The nature of ecological complexity: a protocol for building the narrative. *Ecological Complexity*, **3**: 171–182.
7. Wittgenstein, L. 1973. *Philosophical Investigations*, 3rd edn. New York: Prentice Hall.
8. Zellmer *et al.* The nature of ecological complexity, 178.
9. James. W. 1956 [1896]. *The Will to Believe, Human Immortality, and other Essays on Popular Philosophy*. New York: Dover, 17.
10. James, W. 2006 [1907]. What pragmatism means. In *Pragmatism Old and New: Selected Writings*, ed. S. Haack. Amherst, NY: Prometheus, 294.
11. James, W. 2006 [1906]. Pragmatism's conception of the truth. In *Pragmatism Old and New: Selected Writings,* ed. S. Haack. Amherst, NY: Prometheus, 315.

8

Ecological pragmatism and constrained perspectivism: epistemology

We maintain that knowledge of a particular ecological relationship or system begins with the specification of what is taken to exist (ontology) and its properties (metaphysics). The ecologists' ontological and metaphysical commitments influence what they take themselves to know, for we only investigate properties of that which is taken to possibly exist. From this point, methodologies that accord with epistemic justification, based on that which is taken to be real, can be sensibly developed for a particular problem. For example, whether the ecologist considers a species to be an entity or a process will lead to different types of experimental designs, data collection practices, and analytical approaches. Indeed, epistemological pluralism lies at the heart of our referring to the "practices of ecology" in the title of this book, for it is clear that there is no single, objectively right way to frame ecological questions.

The claim to know something about the subject of study is a matter of receiving feedback in response to actions which derive from our beliefs. This has been called "the pragmatic test," in which the truth of an assertion is validated by its consequences when used.[1] As such, knowing depends on doing. This engagement with the world was explicated in direct terms by John Dewey:[2] "Nature is capable of being understood. But the possibility is realized not by a mind thinking about it from without but by operations conducted from within, operations which give it a new relation summed up in production of a new individual object."

Given the plurality of human needs and desires, a uniquely optimal methodological approach for a broad range of cases is impossible to derive. That is, what may be a highly effective method from one conceptual perspective may not be viable for another. Furthermore, even individual cases may not yield a single, best approach (i.e. being most reasonable and defensible in terms of rational argument) due to incomplete knowledge

106

of constraints. But ecologists need not view a lack of certainty about how to know the world as a liability. From at least one perspective – the political philosophy of John Stuart Mill – epistemic diversity is an asset, as we shall see.

Ecology and politics: pluralism makes strange bedfellows

The methodological pluralism of constrained perspectivism has an important parallel in the realm of politics. John Stuart Mill saw a diversity of ideas as vital to the political process in that a democratic society depends on having a rich and dynamic pool of differing views upon which to draw in solving a constantly changing set of internal and external challenges.[3] Without such a resource, we risk becoming dogmatic and responding to new problems with old ideas that have long since grown stale in the marketplace of ideas. His reasoning applies equally well to science. Elisabeth Lloyd[4] drew the connection between Mill and Paul Feyerabend[5,6] arguing that these philosophers warn us that in neither the political nor scientific realm should we automatically or absolutely reject an approach to knowledge. The applications to ecology are clear. To quash alternative views is to risk missing ideas that are true, unless one is claiming infallibility – a position inimical to science. Moreover, the various methods being proposed or used in ecology would likely each have a partial truth, revealing a portion of that which is sought. That is, unless we claim to already have the whole truth, some element of the alternative may provide a vital piece. In their own ways and times, Mill and Feyerabend went on to contend that even if one's position is entirely correct – or for our purposes, the empirically most successful concept – the idea must be continually and vigorously contested to avoid its becoming mere dogma. This concern is particularly relevant with respect to the education of students, who may come to hold beliefs as a matter of prejudice without comprehension.

Although some have misinterpreted Feyerabend as being anti-scientific because of his defense of witchcraft, astrology, faith healing, and Chinese medicine, Lloyd maintains that his efforts were not intended to undermine, but to strengthen, science. In the same way that democracy is fortified by taking on all comers, Feyerabend viewed science as having the potential to lapse into doctrine reliant on authoritarianism rather than vigorous and continual testing. So, rather than intelligent design being a bad thing for science, this relatively novel assault on evolution served to sharpen the thinking of biologists and clarify the nature of science to society. Likewise, climate change contrarians have served to point the way to where further research was required to better understand Earth's climate system. Feyerabend was, therefore, intentionally attempting to *enact* the precepts of Mill's pluralism.[7] He did not necessarily believe in the alternative positions that he promoted, but he saw them as essential to the integrity of knowledge and the possibility of finding truth:[8] "My use of examples from astrology should not be misunderstood. Astrology bores me to tears. However it was attacked by scientists, Nobel Prize winners among them, without arguments, simply by a show of authority and in this respect deserved a defense."

On pragmatic grounds, however, we would part company with Feyerabend in his assertion that all epistemologies should be given equal rights, equal access to education, and other positions of power. This is a plausible initial position, but pragmatic philosophy (which he seems to endorse) has revealed more or less efficacious methods and perspectives for western society, science and ecology. Given the pragmatists' commitment to fallibilism, perhaps the door should never be shut and locked on alternative views.[9] But in light of limited social resources for human inquiry, and science in particular, we contend that it is unjustified to restage the battle between evolution and creationism, ecology and animism, or any other conflict between alternative ways of knowing the world until and unless: (1) the nonscientific approach has something novel to offer; (2) the problem being solved has changed such that the scientific approach no longer works satisfactorily; or (3) the nature of a new problem is unfamiliar to the extent that it is reasonable to believe that the alternative method warrants new consideration.

We hasten to note that within the field of ecology, the empirical verification of competing accounts of how best to understand natural processes and intervene in ways that accord with our interests are rarely unambiguous. As such, it is incumbent on ecologists not only to tolerate alternative scientific explanations, but to actively pursue and explicitly cultivate such pluralism. For if progress is possible at all, it depends on the critical refinement of ideas.[10]

Basic and applied ways of knowing

The messy, inexact thrashing out of competing methodologies and standards of knowledge within the constraints made explicit by ecologists for a particular problem or class of problems appears to be the fate of a science founded on constrained perspectivism. But we'd suggest that science has done worse in its efforts to know which approach to take (e.g. logical empiricism, Popperian falsification, *ad hominem* attacks, fruitless arguments, and political authoritarianism). Likewise, the marketplace of ideas in a democracy may be inefficient, but there are good reasons to believe that a society can do far worse in its attempts to figure out what to do.

With respect to epistemic issues, pragmatist philosophers would not dismiss purely analytical approaches, but engaging the world is an essential element of attaining knowledge. Dewey, for example, wanted to replace spectatorial idealism (in which the scientist was a spectator of a separate world of pure, logical truths and Platonic forms) with empirical naturalism, grounding knowledge in experiences of the natural world. In effect, the pragmatists viewed knowing as a form of doing. Mental states were, for Dewey, not conditions of knowing directly about the world but procedural instructions for overt action, the consequences of which constitute the object that is said to be known: "In general, we mean by any concept nothing more than a set of operations; *the concept is synonymous with the corresponding set of operations.*"[11] He analogized ideas to tools, such that knowledge and ideas were instruments that could help us deal with a new problem or require modification upon reflection as to the manner in which they were originally designed. Of course, within science a tool can be a theoretical construct, a mathematical equation, a computer simulation, a measuring instrument, an experimental design, or a field exclosure.

Through this pragmatic approach to knowledge, constrained perspectivism also sheds some light on the basic versus applied divide in ecology, a putative chasm into which a great deal of intellectual energy, economic capital, and political effort has been poured.[12,13] Within the philosophical framework we prescribe, this is a false dilemma without substantive value to the field of ecology. This interpretation is based on two important considerations.

First, we take pragmatism to be the position that what is beautiful, right, or true is a matter of whether the actions based on our beliefs genuinely satisfy human needs or wants. As such, we contend that all ecological science is pragmatic; what differs is the particular pragmatic focus of basic and applied research. The basic ecologist casts science in terms of knowledge (or less likely, beauty or morality) – an entirely legitimate human desire. But, the expression "knowledge for knowledge's sake" is vacuous, given that knowledge doesn't have a sake – a condition that could be helped or harmed. The sake of knowledge must be something that can have an interest, such as a human investigator and his colleagues.

Our applied ecologist working in Yellowstone is more self-evidently pragmatic, in that she casts her work in context of a useful outcome, most often in terms of human well-being (e.g. achieving a sustainable harvest of timber or avoiding harmful climate change). This result need not be anthropocentric, but it is unavoidably anthropogenic. That is, she might contend that an ecological study is meant to further the condition of a nonhuman species or an ecological process without direct relevance to the human condition. But just as we are world-makers in terms of reality, we are value-givers in terms of ethics (Box 8.1). So, the applied ecologist attempting to protect the pollination of an orchid that few people will ever see cannot escape having made the pragmatic decision that such a project, at the very least, accords with a sense of well-being – and almost certainly with the interests of some funding source (Box 8.2).

Second, we can understand the basic ecologist as exploring the nature of the constraints in the world (through theoretical probing or empirical testing) – the manner in which reality "pushes back" within a given perspective. The applied ecologist constructs solutions within these constraints, crafting approaches that accord with human interests. In terms of an analogy, consider our interest in having something to sit upon – we desire a chair.[14] The basic scientist investigates the limits of objects (e.g. what are the properties that allow a substance to be made

---

Box 8.1
Leopold, Carson and values

Perhaps the two most influential ecologists with respect to the parallel development of science and values were Aldo Leopold and Rachel Carson. While we have the deepest respect for the contributions of these ecologists, a careful reading of their work leaves us perplexed with respect to their axiological positions.

Leopold's famous contention that "a thing is right when it tends to preserve the integrity, stability, and beauty of the biotic community"

Box 8.1 (cont.)

is hailed as the foundation of modern environmentalism. The role of aesthetics is evident, but Leopold is frustratingly imprecise in explicating what constitutes beauty. Likewise, it is simply not clear what constitutes integrity and stability and – more essentially – why these qualities constitute moral value. He might have chosen diversity, age, naturalness, productivity, or any other of many dozen ecological parameters. Leopold alludes to stability and integrity having instrumental value (a utilitarian ethic), but he also invokes a sense of intrinsic value and a "respect for nature" that seems deontological. Moreover, one can only infer that abiotic entities (e.g. water, air, rocks, and canyons) are at best of instrumental value insofar as they contribute to biotic communities. And finally, why communities, rather than species or individuals, are the loci of value is not entirely evident.

Carson was surely courageous, even heroic, but *Silent Spring* (as well as the *Sea* books) is problematical in that her axiology is implicit and imprecise. It would appear that she believed that natural living entities – particularly individuals, but perhaps also species and communities – had intrinsic value, but this commitment is largely unspoken. Rather, she appeals tangentially to beauty (never making clear what things are beautiful or whether this quality is found in all of nature) and directly to utilitarian concerns. The loss of ecosystem services as well as human health through the misuse of pesticides was the foundation of her moral concern. If so, this might explain why she advocated what strikes a modern ecologist as a rather ill-advised pest management tactic: the introduction of exotic, generalist predators. Perhaps she had not fully appreciated the possible or likely harm to native species or maybe her concern was with sustaining human well-being rather than preserving natural associations.

Our goal in this rather quick and cursory analysis of the profound contributions of Leopold and Carson is not to denigrate their work. Rather, it is to point out that philosophical clarity – as well as scientific rigor and political efficacy – is a virtue worth pursuing. As persuasive as they were, one has to wonder whether environmental ethics and public policy might have been even better served if either of these giants in ecology had provided clear and explicit arguments for their values.

Box 8.2
Basic versus applied science: The pragmatists' approach

The differentiation between basic (theoretical, abstract, mathematical, universal) and applied (empirical, tangible, qualitative, local) science was addressed by the pragmatists in various ways:

Why abstraction is vital to applied science
Artificial simplification or abstraction is a necessary precondition of securing the ability to deal with affairs which are complex ... Abstraction is simply an instance of the economy and efficiency involved in all intelligent practices: deal first with matters that can be effectively handled, and then use the results to go on to cope with more complex affairs.
(John Dewey, The Quest for Certainty, *p. 217)*

Why application is essential to basic science
[Pragmatism] has no objection whatever to the realizing of abstractions, so long as you get about among particulars with their aid and they actually carry you somewhere.
(William James, "What Pragmatism Means" in Pragmatism Old and New, *p. 303)*

Why applied science has been disparaged
[The] exaltation of pure intellect and its activity above practical affairs is fundamentally connected with the quest for a certainty which shall be absolute and unshakeable. The distinctive characteristic of practical activity, one of which is so inherent that it cannot be eliminated, is the uncertainty which attends it.
(John Dewey, The Quest for Certainty, *p. 6)*

Why the basic-applied conflict is vacuous
[A]ction is at the heart of ideas. The experimental practice of knowing, when taken to supply the pattern of philosophic doctrine of mind and its organs, eliminates the age-old separation of theory and practice.
It discloses that knowing is itself a kind of action, the only one which progressively and securely clothes natural existence with realized meanings.
(John Dewey, The Quest for Certainty, *p. 167)*

---

into a chair; what are the bounds of structures that can be chairs?), while the applied scientist uses this knowledge to make chairs (e.g. having been told that water can't be used, does not attempt to make a chair out of water). But the relationships are not unidirectional. Perhaps the applied scientist doubts the basic scientists' claim about the suitability of water, so makes a chair out of ice. This development requires that the basic scientist expand or clarify the boundaries that have been ascribed to reality. This dialectical process between theory and practice allows the scientists to refine both the theory and construction of chairs. If we substitute "forests" and their properties (or "wetlands," "conservation reserves," "species re-introductions," etc.) for "chairs" and their qualities, the relevance of the analogy to ecology should be evident.

Contingent knowing is genuine knowledge

Although constrained perspectivism makes the development of general ecological theories less likely – or at least less important to the validity of our science – it is not a formula for radical nominalism (the contention that we can only know about individual cases). Rather, we suggest that within an ontological and metaphysical framework, a general understanding of cases that resemble one another in relevant ways may be possible. Even across frameworks, we may be able to claim knowledge of sufficiently similar entities, processes, and properties. And once again, progress depends on a dialectic; in this context the exchange is between how ecologists shape reality (ontology/metaphysics) and how we claim to know (epistemology).

Ecology is a highly inductive science, such that we typically begin with a set of particular cases which permit the exploration of general patterns, explanations, and predictions. Solutions that work by whatever pragmatic standard has been chosen for specific instances can be extended to novel cases and the results used to refine the framework by expanding or reducing the presumptive constraints in accordance with the outcome. For example, if an ecological explanation accounts for some phenomenon on temperate grasslands and is then used to predict the results of a study on a tropical grassland, we might find that the outcome fits the expected results. If so, then the system to which our knowledge obtains is expanded to provisionally include all grasslands, and "grasslands" become a natural kind insofar as our explanation is concerned. However, if the prediction fails, then temperate and tropical grasslands are different kinds of systems with respect to our knowledge, vis-à-vis the explanation. In this way, useful generalizations are built through the interplay between accommodation and prediction of cases which ecologists pragmatically classify – as exemplified by the cases that we explored in Chapter 2.

Central to the process of coming to know about the world in ecological terms is a recognition that at some scale there will be particular conditions which are fundamentally irreproducible (no two cases are precisely identical). The task of the ecologist is to select an appropriate scale which provides a context for developing operational diagnostics that allow individual cases to be aggregated with respect to entities, processes, and properties relevant to human concerns. In simple terms, there simply are no ecological problems with respect to human interests if we take a sufficiently short (one minute) or long (one billion years) perspective. In more practical terms for ecologists, using 1-m^2 cages to determine mortality factors in an insect population will include competitive processes, exclude many predatory processes (depending on the size of the mesh), and exclude all large-scale epizootic processes (or, conversely, promote diseases within confined areas) so the explanations provided through such experiments can probably not be aggregated to scales of 1 km^2. As such, by establishing the bounds of a framework, scale plays a critical role in establishing what will constitute evidence confirming or refuting a hypothesis.[15] As Wimsatt put it, "[O]ur concept of evidence would predictably be much more contingent, contextual, and historicist – complex, relational, and sensitive to context, but that is not to say relativistic."[16]

Understanding the importance of scale to epistemology in constrained perspectivism may also allow ecologists to escape from the ideal of an unconditional, epistemic standard. That is, the ecologist must consciously choose a standard of justification and evidence that accords with the nature of the problem as understood from an explicitly identified perspective and a clear expression of interests. As such, the question of whether one should ascribe to a particular epistemic standard of justification (e.g. strict falsification or weight of evidence) cannot be given a decontextualized answer.

In everyday terms, we have different epistemic standards for different domains of life. That is to say, we might have a rather low standard for issues such as whether to change our beliefs about brands of ketchup and much higher standards for changing our beliefs about types of medical procedures. With respect to science, the same conditionality seems appropriate. A Popperian approach may be a serious handicap when one is faced with a pressing ecological problem that demands a solution for which we can't afford to use strict falsification (which philosophers of science have largely abandoned) because any model developed under conditions of limited data, time, and resources will almost necessarily be wrong.[17,18] We seek an answer that is "close enough." On the other hand, we might have a more rigorous epistemic standard concerning a particular paleoecological problem for which there is no immediate harm to human well-being if we demand a high level of evidence before making a claim of knowledge.

So, how does constrained perspectivism fare, by its own standards, as a descriptive and prescriptive account of the philosophy of ecology? The problem we are attempting to solve is largely that of ecology having either no coherent philosophical structure or adopting that of the physical sciences. In either case, the result often takes the form of fruitless arguments, squandered resources (material and intellectual), and unrealized potentials with regard to solving serious problems of humans and other beings. Constrained perspectivism is almost certainly a "partial truth" with regard to how ecologists do, and ought to, act. But we are not so much concerned with what ecology "really" is in some strangely objective or absolute way – rather, we are hoping that the adoption of constrained perspectivism will work. That is to say, this philosophical framework will provide a means by which ecologists can become more productive, creative, and ultimately effective. To be consistent, therefore, we encourage the reader to ask not, "Is constrained perspectivism a universally correct account of ecological science?" but rather, "Under what conditions or within which domains is constrained perspectivism useful?"

Endnotes

1. Schiller, F. C. S. 2006 [1907]. The making of truth. In *Pragmatism Old and New: Selected Writings,* ed. S. Haack. Amherst, NY: Prometheus, 49.
2. Dewey, J. 1960 [1929]. *The Quest for Certainty: A Study of the Relation of Knowledge and Action.* New York: Capricorn, 215.
3. Mill, J. S. 1977 [1859]. *Essays on Politics and Society, Collected Works of John Stuart Mill,* vol. 18. Toronto: University of Toronto Press.
4. Lloyd, E. A. Feyerabend, Mill, and pluralism. *Philosophy of Science,* **64**: 396–407.
5. Feyerabend, P. K. 1975. *Against Method: Outline of an Anarchistic Theory of Knowledge.* London: Verso.
6. Feyerabend, P. K. 1978. *Science in a Free Society.* London: Verso.
7. Lloyd, Feyerabend, Mill, and pluralism.
8. Feyerabend, P. K. 1991. *Three Dialogues on Knowledge.* Oxford: Blackwell, 165.
9. Fallibilism, poorly understood, can be wielded as a political weapon. In *Doubt is Their Product: How Industry's Assault on Science Threatens your Health* (New York: Oxford University Press), David Michael shows how the "product defense industry" uses the uncertainty of science to confuse people about the risks of tobacco, environmental toxins, global warming, and just about anything else where fallibility can be converted into inaction. The fallibilist recognizes that he or she may be wrong, but this is a very different stance than the assertion that all claims are equally prone to error. That we do not know everything is rather noncontroversial, but that we do not know anything is a

kind of degenerate fallibilism that entrenches the status quo in that
uncertainty is used disingenuously by those who wish to persist in an activity
that has evident – but not certain – risk.

10. Feyerabend maintained that science could lead towards truth, rather than
being an aimless wander within the space of possibilities or the passing of
views from one authoritative class to another. He seems to have had in mind
a kind of correspondent truth, although the claim would hold equally for
a pragmatic understanding of truth.

11. Dewey, *The Quest for Certainty*, 111.

12. Costanza, R. 1993. Developing ecological research that is relevant to
achieving sustainability. *Ecological Applications*, **3**: 579–581.

13. Ludwig, D. 1993. Environmental sustainability: magic, science, and religion
in natural resource management. *Ecological Applications*, **3**: 547–549.

14. Schiller, The making of truth, p. 499.

15. Pickett, S. T. A., J. Kolasa and C. G. Jones. 2007. *Ecological Understanding:
The Nature of Theory and the Theory of Nature*. San Diego: Academic Press.

16. Wimsatt, W. C. 2007. *Re-Engineering Philosophy for Limited Beings: Piecewise
Approximations to Reality*. Cambridge, MA: Harvard University Press, 157.

17. Hilborn, R. and M. Mangel. 1997. *The Ecological Detective: Confronting
Models with Data*. Princeton: Princeton University Press.

18. Pickett *et al. Ecological Understanding*.

9

Ecological pragmatism and constrained perspectivism: a summary

World-making through philosophical and ecological collaboration

Although philosophers sometimes make sweeping and pretentious claims of knowledge, the authentic task of philosophy is to mediate the conversation among various ways of knowing, such that a critically constructive dialogue emerges (Box 9.1). As John Dewey argued:[1]

> The situation defines the vital office of present philosophy. It has to search out and disclose the obstructions; to criticize the habits of the mind which stand in the way; to focus reflection upon needs congruous to present life; to interpret the conclusions of science with respect to their consequences for our beliefs about purposes and values in all phases of life.

Constrained perspectivism can serve as a liaison among ecological views. By drawing ecologists back to the pragmatic context, it may be possible to mitigate fruitless arguments and facilitate productive exchanges. This philosophical framework might at least compel ecologists to recognize that there is no such thing as a view from nowhere (to paraphrase Thomas Nagel) – all scientific claims are perspectival. Perhaps the best way of capturing this necessary condition is by considering art. That is, all paintings and photographs are taken from somewhere and the location of the artist defines the perspective. M. C. Escher's work is particularly effective in drawing our attention to the inevitability of perspective – sometimes several perspectives are included but even Escher could not draw a view from nowhere (Box 9.2).[2]

However, this is not to say that we, like William Wimsatt[3], are describing or advocating a freewheeling, "anything goes" relativism of the sort that might be suggested in the extreme versions of postmodernism. While

117

Box 9.1
Why philosophy and science need one another

In considering the relationship between philosophy and science,
we might go so far as to suggest substituting "ecology/ecologists"
for "philosophy/philosophers" in John Dewey's critical essay,
"The need for a recovery of philosophy":

> Philosophy recovers itself when it ceases to be a device for dealing with
> the problems of philosophers and becomes a method, cultivated by
> philosophers, for dealing with the problems of men.

One might also substitute "people" for "men," but perhaps "life
on earth," or some other entity or process might be even more
compelling depending on one's axiology. Ecology and philosophy
offer one another alternative perspectives from which each may see
itself more clearly. For if human knowledge is relational, then all
fields are in service to one another. This mutualism is evident in
Dewey's understanding of the ways in which pragmatism and
science might goad one another to more rigorous thought and
effective action: ("School conditions and the training of thought,"
in *Pragmatism: Old and New*, p. 350):

> The pragmatic theory thus claims faithfully to represent the spirit, that is
> the method, of science, which (1) regards all statements as provisional or
> hypothetical till submitted to experimental test; (2) endeavors to frame its
> statements in terms which will themselves indicate the procedures required
> to test them; and (3) never forgets that even its assured propositions are
> but the summaries of prior inquiries and testings, and therefore subject to
> any revision demanded by further inquiries.

Ronald Giere's view of science renounces a degenerate relativism based
on intersubjective objectivity of visual perspectives,[4] we are making both
a somewhat different and more expansive claim. That is, we are concerned not only or primarily with the origin of differences in visual
reports within science, but with sensory and conceptual perspectives.
And we maintain that the constraint on this broad range of perspectives
is ultimately external to the observers, although intersubjective agreement
is surely an important and interesting criterion for forming beliefs about
the world (Box 9.3).

Box 9.2
Perspectivism: a model

M. C. Escher produced many works that reveal the world through multiple perspectives – all at the same time. The lithograph shown here (*Relativity*, 1953) depicts people engaged in one of three domains of gravity, each being orthogonal to the others. Each of the stairways connect two of the domains, such that they provide an interface or point of communication between different perspectives. It is evident that there is no "right" perspective – but there are useful ones and wrong ones. An individual who chose to act in the context of another domain would seemingly fall headlong into the image. So, there is more than one way to be right, but one can also be mistaken from within a domain.

M. C. Escher's *Relativity* ©2008 The M.C. Escher Company-Holland. All rights reserved. www.mcescher.com

---

Box 9.3
The pragmatists' view of science

Pragmatists are not unified in their understanding of science, but a set of common themes emerges in terms of the qualities of science.

Science is contingent: William James saw science as a series of episodic solutions or "flights and perchings" which presaged Thomas Kuhn's *Structure of Scientific Revolutions*. For James, there is no grand unification or teleological progress toward absolute truth; there are only solutions that work for the time-being and must change as new problems arise – often from the earlier solutions themselves.

Science is perspectival: Nelson Goodman argued that we discover what we are prepared to find, that we are "blind to that which neither helps nor hinders our pursuits." The scientist rejects or idealizes away those aspects of the world deemed irrelevant to the concept of interest, "while generating quantities of filling for curves suggested by sparse data, and erecting elaborate structures on the basis of meager observation." ("Words, works, worlds," in *Pragmatism Old and New*, p. 611)

Science is instrumentalist: John Dewey understood theories as tools for exploring the world. In defending science from the criticism that it was constantly being revised, Dewey argued, "No one would dream of reflecting adversely upon the evolution of mechanical inventions because ... the mechanized tractor substituted for the horse-drawn mower. We are obviously confronted with betterment of the instrumentalities that are employed to secure consequences." (*The Quest for Certainty*, p. 192)

Science is justified knowing: Charles Peirce viewed science as a way of reasoning responsibly in the light of logic and evidence, while James noted that various sciences have different ways of obtaining reliable knowledge. Dewey went further to argue that science had no privileged relation to reality for, depending on one's interests, "the painter may know colors as well as the physicist ... the farmer may know soils and plants as truly as the botanist and mineralogist ... There are as many conceptions of knowledge as there are distinctive operations by which problematic situations are resolved." (*The Quest for Certainty*, p. 192)

Box 9.3 (cont.)

Science is pragmatic practice: James maintained that science and pragmatism did not stand for any particular results but were simply methods. According to Dewey, the approaches entailed empirical verification and openness to continued testing and revision. In short, James saw scientific pragmatism as a kind of orientation reflected in, "The attitude of looking away from first things, principles, 'categories,' supposed necessities; and of looking towards last things, fruits, consequences, facts." (*Pragmatism Old and New*, p. 295)

Constrained perspectivism is consistent with the pragmatists who see reality as being negotiated via subjective interests working within the constraints imposed by objective reality. This type of world-making was propounded by philosophers such as Nelson Goodman and F.C.S. Schiller.[5] In this regard, we can expand on Schiller's classic comparison of reality and chairs, but offer an example more relevant to ecology.

Let us suppose that our Yellowstone ecologist wants to get a closer look at the bison. A spotting scope might work, but she decides that an observation blind would really do the trick. She does not wander across the meadow and attempt to "discover" a blind amidst the trees, rocks, and streams. Nor does she hope to "create" a blind out of her imagination. Rather, she "makes" a blind by engaging the stuff of the meadow and crafting it to meet her needs. However, not just anything in the meadow can be made into a blind; the world pushes back against her interests if she attempts to construct a blind out of water or pine needles. Rather, she quite sensibly fashions a frame out of willow branches (aspen branches would also work) and covers it with grasses (or a camouflage tarp that accords with the color perception and visual acuity of bison).

For Schiller and other pragmatists, our world is made in an analogous manner, we craft from objective potential a version of reality that accords with our interests (including our desire for scientific understanding). The world that is constructed from the raw material of mind-independent existence can be pushed to the bounds of its objective properties until it pushes back and constrains our world-making project. In terms of ecology, we would suggest that the bounds of objective reality are far broader than the constraints imposed by scientific or cultural convention. In Kuhn's terms, scientific paradigms set quasi-metaphysical limits based on implicit social agreement rather than actual boundaries.[6]

We hasten to point out that our understanding of pluralism (as with that of Mitchell and Dietrich[7]) is that this approach is the endpoint of the ecological enterprise, not merely a means to the ends of grand unification, as suggested by Kitcher.[8] He argued that limited evidence made it prudent to sustain multiple, competing hypotheses, while the scientific community's ultimate goal, "is to arrive at universal acceptance of the true theory." This view reflects a version of pragmatism which holds that through continual, empirical verification we will converge on objective truth, although this may be, in the end, a domain-specific or perspectival truth.[9] We reject the notion that there is a single, true perspective toward which we can or do converge – a position that resonates with the views of ecologists who remain dubious of the search for grand unification (see Chapter 2). We note, however, that there may well be epistemological convergence on the methods or practices that are best suited for understanding a particular problem.

"But then," one might object, "what happens to truth?" That is, the ecologist may be legitimately concerned about whether pragmatism will support claims about the natural world that are within the bounds of biophysical possibility but outside the accepted scientific norms. While constrained perspectivism may well permit in a broader range of concepts than traditional limits allow, reality will effectively filter out good science (that which accords with objective reality and our subjective interests) from wishful thinking. The communal or pluralistic process of science practiced in a manner consistent with constrained perspectivism can also be taken as a means of avoiding – or at least not perpetuating – erroneous interpretations of the world. Wimsatt captures the epistemic power of pluralism through the notion of "robustness analysis" by which he means substantiation of an idea by more than one means (or, we might say, multiple perspectives):[10]

> One may ask whether any set of such diverse activities, as would fit all these items (and as exemplified in the expanded discussion below), is usefully combined under the umbrella term *robustness analysis* [author's italics].
> I believe that the answer must be yes, for two reasons. First, all the variants and uses of robustness have a common theme in the distinguishing of the real from the illusory; the reliable from the unreliable; the objective from the subjective; the object of focus from artifacts of perspective; and, in general, that which is regarded as ontologically and epistemologically trustworthy and valuable from that which is unreliable, ungeneralizable, worthless, and fleeting.

But the truth will not be granted by supposed correspondence to a universal, objective reality. Rather, as William James eloquently argued,

---

Box 9.4
The evolution of truth

A pragmatist views birds' wings and scientific concepts as instruments that work (or fail) in the world with respect to the interests of those possessing them. And William Wimsatt argued that ideas are the products of evolutionary processes (2007. *Re-Engineering Philosophy for Limited Beings*, p. 135):

> New systems that facilitate mechanisms by which some elements can come to play a generative or foundational role relative to others are always pivotal innovations in the history of evolution, as well as – much more recently – in the history of ideas. Mathematics, foundational theories, generative grammars, and computer programs attract attention as particularly powerful ways of organizing and producing complex knowledge structures and systems of behavior. They not only produce or accumulate downstream products, but they do so systematically and relatively easily. *Once they appear, generative systems become pivotal in any world where evolution is possible*: biological, psychological, scientific, technological, or cultural. Generative systems come to dominate in evolution, and are rapidly retuned and refined for increasing efficiency, replication rate, and fidelity, as soon as they are invented. We must suppose that even modest improvements in them spread like wildfire.

---

truth becomes an instrument of successful action rather than a conceptionless reflection or mirror of existence. An idea comes to be true in a manner not unlike the way a trait is said to be fit in evolutionary terms – it succeeds in the context of its environment and in light of the problem that it solves better than the alternatives. The analogy that the pragmatists drew between the survival and modification of ideas and the evolution of species has been recently highlighted in the philosophy of science (Box 9.4). For James, truth is something that happens to an idea through the world responding to actions that are based on the idea. John Dewey had a parallel notion, arguing that knowledge arises from an organism's actions, which allow prediction and control of its world.[11]

As such, the pragmatic foundation of science addresses a critical issue that Giere did not fully explicate in his account of scientific perspectivism. In reference to the life sciences, he argued that, "some specific evolutionary models structured according to evolutionary principles have been *successfully* applied to real populations [emphasis added],"[12] and "In the case of finches, for example, a rough positive correlation between strength of beak and hardness of seeds is *enough* [emphasis added]."[13] It is not clear, however, how Giere would have the scientist know what constitutes "success" or "enough." These critical qualifiers to an understanding of science via perspectivism are clarified by the pragmatic standard. That is to say, a model is successful or a correlation is good enough in terms of the interests that motivate the inquiry. A predator-prey model may be successful with respect to the desires of a basic researcher seeking an aesthetically pleasing account of dynamics within a chemostat, but not good enough for the applied scientist attempting to develop a wolf management policy for a national park.

Constrained perspectivism and the pragmatic tradition

The ontological and metaphysical foundations of constrained perspectivism accord with the essential elements of modern pragmatism.[14] And, we would contend, this philosophical tradition provides both valuable descriptive and prescriptive force to the scientific venture. In particular, the pragmatist is committed to (Box 9.5):

- anti-foundationalism (there is not an unconditional, unitary basis for claims of beauty, right, or truth),
- fallibilism (the presumption that one's claims are only partially correct and open to revision, with the consequent admonition that given the multiplicative nature of probability: "Fallible thinkers should avoid long serial chains of reasoning"[15]),
- critical community of inquiry (to progress in understanding and effectiveness, one must be embedded within a group of investigators who authentically and constructively engage in mutual doubt),
- dynamic contingency (an appreciation that claims of knowing are subject to revision as the conditions under which they have been developed change), and
- pluralism (the understanding that valid claims of beauty, truth, or right may be irreducible, incommensurable, and irresolvable).

The pluralism implicit in constrained perspectivism is particularly important, as it suggests that there is not a single, correct context from which to conduct ecological investigations. The pluralistic approach

---

Box 9.5
Fallibilism and pluralism

The two central tenets of pragmatism are that we may be wrong and that there is more than one way to be right.

Fallibilism, John Dewey argued, is the position that no knowledge is assured because everything we know is the product of particular acts of inquiry which are necessarily prone to errors of perception and interpretation. William James maintained that being wrong was an inherent risk of empiricism given that our experiences and experiments are always limited:

> Believe nothing, [the skeptic] tells us, keep your mind in suspense forever, rather than by closing in on insufficient evidence incur the awful risk of believing lies. It is like a general informing his soldiers that it is better to keep out of battle forever than to risk a single wound. Not so are victories either over enemy or nature gained. Our errors are surely not such awful solemn things. In a world where we are so certain to incur them in spite of all our caution, a certain lightness of heart seems healthier than this excessive nervousness on their behalf
>
> (*"The will to believe,"* in Pragmatism Old and New, *p. 235)*

Pluralism, according to Nelson Goodman, admits that various fields – such as ecology or psychology – may have different, but no less rigorous, standards than physics. There is no presumption of reducibility to a single frame of reference, but such contextualism does not degenerate into relativism:

> That right versions and actual worlds are many does not obliterate the distinction between right and wrong versions, does not recognize merely possible worlds answering to wrong versions, and does not imply that all right alternatives are equally good for every or indeed for any purpose . . . Moreover, while readiness to recognize alternative worlds may be liberating, and suggestive of new avenues of exploration, a willingness to welcome all worlds builds none . . . A broad mind is no substitute for hard work.
>
> (*"Words, works, worlds,"* in Pragmatism Old and New, *p. 616)*

---

opens the door to there being more than one truth about the world. This position falls between a degenerate or nihilistic relativism contending that there are an unlimited number of truths (or functionally no truth) versus a hard-driving realism arguing for a single, objective truth. As with pragmatism, our constrained perspectivism holds that there are multiple, contingent truths – and that there are ways to be wrong (e.g. attempting to make chairs or blinds out of pine needles). Such a position is similar to Hillary Putnam's internal realism, a pragmatic view which posits that truth is relative to a conceptual scheme such that a scientific claim can be true (or false) within its domain but cannot be taken to be a whole or universal truth.[16] However, we would differ from Putnam in that he also holds that scientific truths have no necessary correspondence with a mind-independent reality (which, for him, may not even exist). We would contend that scientific assertions will be ultimately constrained by feed-back from reality – a lack of effective engagement with some particular, but relevant, aspect of the world.

The consequences of constrained perspectivism for ecology

The results of adopting the ontology, metaphysics, and epistemology of constrained perspectivism are not trivial, but neither are the consequences radically unfamiliar to ecologists. That is, our proposed framework may be descriptive of how many researchers already think about science and will hopefully serve our colleagues by making their implicit understanding philosophically explicit (Box 9.6). At the same time, it is prescriptive in calling for an extension of this approach in ways that ecologists may not currently practice. In particular, we would point out four important outcomes of constrained perspectivism.

First, we take the view that although the world (entities and processes) and its properties are not infinitely plastic, they are extremely malleable. Perspectivism is constrained by the nature of objective reality, but these limits also may be contextual. That is, to extend Schiller's analogy, whether or not a chair can be made out of water depends on the tempera-ture (ice can be carved). There are many worlds that the ecologist can craft from the raw material of actual existence. To follow this line of development to its logical conclusion, we would suggest that there are far more possible and potentially useful (with respect to the enormous range of human interests) chairs in the furniture shop of ecology than might be generally believed or imagined. Such a situation demands a high tolerance for unfamiliar and unconventional chairs, requires a healthy skepticism of new chair designs, and eschews the singular devotion of resources to sanding and buffing the same old chairs.

Box 9.6
Pragmatists' views of perspectivism

In "The making of truth," F. C. S. Schiller captured the nature of perspectivism in familiar, everyday terms:

> For every judgment is essentially an experiment, which, to be tested, must be acted on. If it is really true that "this" is a chair, it can be sat in. If it is a hallucination, it cannot. If it is broken, it is not a chair in the sense my interest demanded. For I made the judgment under the prompting of a desire to sit ... Of course, however, if my interest was not that of a mere sitter, but of a collector or dealer in ancient furniture, my first judgment may have been woefully inadequate, and may need to be revised. "Success," therefore, in validating a "truth," is a relative term, *relative to the purpose* with which the truth was claimed. The "same" predication may be "true" for me and "false" for you, if our purposes are different. As for a truth in the abstract, and relative to no purpose, it is plainly unmeaning. For it never becomes even a claim, and is never tested, and cannot, therefore be validated.
>
> *(Pragmatism Old and New, p. 499)*

In "Words, works, worlds," Nelson Goodman elucidated the implications of perspectivism in a scientific context:

> Consider, to begin with, the statements "The sun always moves" and "The sun never moves" which, although equally true, are at odds with each other. Shall we say, then, that they describe different worlds, and indeed that there are as many different worlds as there are such mutually exclusive truths? Rather, we are inclined to regard the two strings of words not as complete statements with truth-values of their own but as elliptical for some such statement as "Under frame of reference *A*, the sun always moves," and "Under frame of reference *B*, the sun never moves" – statements that may both be true for the same world ... If I ask about the world, you can offer to tell me how it is under one or more frames of reference; but if I insist that you tell me how it is apart from all frames, what can you say? We are confined to ways of describing whatever is described.
>
> *(Pragmatism Old and New, pp. 601–602)*

Second, given both the richness and malleability of the world, eco-logical reality is fundamentally contextual and contingent (e.g. depending on the spatiotemporal scale, measurement precision, and relationships of interest to the researcher). Ecological change can be cast in terms of persistent, nonequilibrial dynamics; multiple, unstable equilibria; or single, stable equilibria. As world-makers, ecologists must be keenly aware, intentionally discerning, transparently explicit, and fully responsible in selecting their ontologies and metaphysics. Vague and imprecise philosophy will feed the unnecessary conflict and confusion that too often consumes the time, energy, and resources of ecology.

Third, constrained perspectivism holds that the ontological commitments and metaphysical positions of ecologists should be based on pragmatic considerations. By this we mean to invoke the full range of human interests, including our need for food, shelter, energy, a livable environment, a functional social system, and physical and mental health – as well as our desire for knowledge, understanding, comfort, material wealth, beauty, and meaning. In addition to these overarching concerns, ecologists must often engage specific considerations of immediate problems, including the practicality of gathering relevant data, availability of relevant theoretical structures, spatiotemporal context, and histories of the entities and processes. And finally, the ecologist's pragmatic considerations will likely include cultural, social, economic, political, and legal factors that make an investigation socially acceptable, publicly or privately fundable, and ultimately publishable. Indeed, we would readily admit that the ideas in this book were chosen, shaped, and presented in light of pragmatic considerations – our argument for constrained perspectivism cannot escape its own implications. And it is for this reason that we must further part company with the otherwise compelling ecological philosophy of Zellmer *et al.*[17] For their metaphysics of individual relativism must presumably constitute just another story, no more or less true than any other narrative and therefore quite impossible to argue for or against, except to the extent that one happens to find it subjectively appealing.

Fourth, an essential task of the ecologist is to establish ontological commitments and invoke tangible properties that facilitate scientific inquiry. For example, an ecologist working with a living system that is undergoing change such that it must be understood in historical terms might choose a process-based ontology (e.g. the temporal, Heraclitian perspective) rather than a material-based ontology (e.g. the static, Platonic perspective). The point being that what is taken to be real for our purposes is a choice that we make within the broad constraints of the world. With this realization comes awareness of, and responsibility for, our subjective decisions. At this level of science (as opposed to the design of experiments, collection of data, and analysis of results), subjectivity is not an insidious or avoidable factor. Subjective conceptualization of entities, processes, and properties is only problematical when ecologists do so without being conscious of their perspective, a situation which leads to dogmatism. And, to the extent that such lack of awareness limits a scientist's access to concepts relevant to solving serious problems that harm individuals or society, this is also a matter of ethical concern.

In this explication of constrained perspectivism, we have attempted to provide examples to illustrate various concepts and aspects (Box 9.2). However, our philosophical approach is not restricted to imaginative applications. There are real-world ecological studies which exemplify many of the tenets of constrained perspectivism and demonstrate the viability of our conceptual framework.

Endnotes

1. Dewey, J. 1960 [1929]. *The Quest for Certainty: A Study of the Relation of Knowledge and Action*. New York: Capricorn, 313.
2. Nagel, T. 1989. *The View from Nowhere*. New York: Oxford University Press.
3. Wimsatt, W. C. 2007. *Re-Engineering Philosophy for Limited Beings: Piecewise Approximations to Reality*. Cambridge, MA: Harvard University Press, 49.
4. Giere, R. N. 2006. *Scientific Perspectivism*. Chicago: University of Chicago Press.
5. Haack, S., ed. 2006. *Pragmatism Old and New: Selected Writings*. Amherst, NY: Prometheus.
6. Kuhn, T. S. 1996. *The Structure of Scientific Revolutions*. Chicago: University of Chicago Press.
7. Mitchell, S. and M. R. Dietrich. 2006. Integration without unification: An argument for pluralism in the biological sciences. *American Naturalist*, **168**: 73–79.
8. Kitcher, P. 1990. The division of cognitive labor. *Journal of Philosophy*, **87**: 5–22.
9. Macbeth, D. 2007. Pragmatism and Objective Truth. In *New Pragmatists*, ed. C. Misak. New York: Oxford University Press, 169–192.
10. Wimsatt, *Re-Engineering Philosophy for Limited Beings*, 46.
11. Alexander, T. M. 1987. *John Dewey's Theory of Art, Experience, and Nature: The Horizons of Feeling*. Albany: State University of New York Press.
12. Giere, *Scientific Perspectivism*, 69.
13. Giere, *Scientific Perspectivism*, 72.
14. Bernstein, R. J. 1997 [1988]. Pragmatism, Pluralism, and the healing of wounds, In *Pragmatism, A Reader*, ed. L. Menand. New York: Vintage, 382–401.
15. Wimsatt, *Re-Engineering Philosophy for Limited Beings*, 49.
16. Putnam, H. W. 1987. *The Many Faces of Realism*. La Salle, IL: Open Court.
17. Zellmer, A. J., T. F. H. Allen and K. Kesseboehmer. 2006. The nature of ecological complexity: a protocol for building the narrative. *Ecological Complexity*, **3**: 171–182.

10

The practice of constrained perspectivism in ecology

How do ecologists practice philosophy?

Constrained perspectivism *prescribes* a pluralistic and pragmatic acceptance of multiple ontologies, metaphysics and epistemologies necessary to understand those slices of nature of interest to ecology. Based on the authors' experience, constrained perspectivism turns out to roughly *describe* the widespread flexibility ecologists have long practiced in the pursuit of scientific understanding. Contrary to unified philosophical stances portrayed in textbooks and classrooms, we contend that most ecologists move readily (but not blithely) among ontologies and associated metaphysical properties to achieve their current purposes, that they select and employ those theoretical insights that hold promise for particular, novel questions, and that they accept implicitly (if not explicitly) the boundary conditions or contingencies that limit ecological theories in their application.[1] In effect, ecologists function as scientific anti-realists who take an instrumentalist approach in which a perspective – including its terminological, theoretical, and methodological elements – is judged as true by virtue of capacity to guide the investigator's actions so as to produce outcomes according with his or her interests.

Indeed, instrumentalism is perhaps the most *apropos* term for our philosophical approach insofar as constrained perspectivism can be seen as a tool of science, or perhaps still better as a bridge.[2] That is, much of ecology involves the utilization of general theory to particular cases by constraint of that theory to contingencies. Perhaps a pragmatic epistemology would be acceptance of contingencies as the practical bridge from the highly general theories of rationalist and realist origins to the applications in particular cases. In other words, we might adopt as a pragmatic epistemology the assumption that not only must domains be declared at the outset,[3] but contingencies must be specified when adopting a general theory for particular classes of cases, or for individual cases. Schrader-Frechette and McCoy[4] describe this approach for a number of special cases in conservation biology.

In this chapter, we present evidence for how constrained perspectivism not only prescribes, but also describes how ecologists often operate, even if unconsciously. We contend that ecologists address questions much more like medical doctors diagnosing symptoms and prescribing treatments, than like classical scientists consciously developing hypotheses and testing them by standard epistemological methods. As there are many ways to practice medicine (e.g. psychiatry, dermatology, gerontology, and pediatric orthopedics), there are also many ways to practice ecology. Our common interests, goals, knowledge, and skills are manifest in the varied practices of ecology. To a certain extent, the results of an ecological study may become dressed up in more formal terms. This seems to work for us, as long as we have a good tool kit of theories to guide us along with the knowledge and data to apply to the theories. William Wimsatt[5] claims that scientists work through heuristic mental processes that are not what we usually consider "reasoning" (Box 10.1). We think he would agree, though, that these are not naïve heuristics, but are acquired through training and experience. And, inevitably, some scientists develop better heuristics as well as a greater capacity to correct them than do others. We are in basic concordance with Wimsatt, but prefer to analyze the process in more philosophically formal and less psychologically mysterious terms. In particular, we examine how ecologists have pragmatically navigated between rationalist general theory (with its troubling elements of Platonic idealism) and realist theory (with its dual problems of unlimited contingencies and nominalism).

General theory and pragmatism in population ecology

The population level of organization is perhaps the best known in ecology in terms of characterizing nature through rationalism. For example, the competition and predator-prey models introduced by Lotka and Volterra have been idealized in textbooks for over 50 years. But those at the rationalist end of the continuum, along with ecologists of an empirical bent, recognized the necessity of contingencies. Examples include terms for immigration-emigration,[6] environmental heterogeneities,[7] and historical circumstances, especially starting conditions.[8] As the need for more

---

Box 10.1
Flowing with our heuristic impulses: Zen science

William Wimsatt describes scientific problem-solving as a much
more informal process than many scientists might see their practices.
The following quotes illustrate his argument.

> In the real world, knowledge of how to use or test a theory does not come
> packaged with its axioms! We who are surrounded by technical systems
> often forget or underestimate the learned wizardry embedded in the
> knowledge and practices of those who work on them – heuristic
> knowledge of breakdowns; their likely causes; and how to find, fix, and
> prevent them. Auto mechanics, doctors, engineers, programmers, and
> other students of mechanisms learn how to debug their preferred systems
> by localizing and fixing the faults that occur; both in their machines and
> in their procedures for working on them. *This works for the mind no less
> than for any of our other tools.*
>
> *(p. 22, author's italics)*
>
> We are not (omnipotent) LaPlacean demons, and any image of science
> that tells us how to behave as if we were still fails to give useful guidance
> for real scientists in the real world ... A more realistic model of the
> scientist as problem solver and decision maker includes the existence of
> such limitations and is capable of providing real guidance and a better fit
> with actual practice ... A central feature of it is the use of "cost-effective"
> heuristic procedures for collecting data, simplifying problems, and
> generating solutions.
>
> *(p. 78)*
>
> There is a mythology among philosophers of science ... that this
> (*piecemeal engineering*) cannot be done, that a theory or model meets its
> experimental tests wholesale and must be taken or rejected as a whole.
> Not only science, but also technology and evolution would be impossible if
> this were true in this and in logically similar cases. That this thesis is false
> is demonstrated daily by scientists in their labs and studies, who modify
> experimental designs, models, and theories piecemeal.
>
> *(p. 103)*

---

contingencies arose, functional boundaries, performance constraints,
along with additional variables and locally appropriate parameter values
all became wrapped around the core theory for making predictions or
providing explanations at a resolution necessary for practical purposes in
the context of particular cases.[9]

In his analysis of the existence of ecological laws, Lawton[10] explored
the necessity of contingencies at the population level, concluding:

The theory of population dynamics, the search for ecological rules, is contingent on the organism and its environment. It is doubtful that this theory will ever be genuinely predictive, in the sense that given a species name, or type of organism, and where it lives, ecologists could, with any degree of certainty, specify the kind(s) of population dynamics it will display without actually having seen a time series, or knowing anything else about its biology. But all the evidence suggests that the contingent theory isn't so complex and multidimensional as to make the range of population dynamics shown by organisms perfectly understandable in terms of a set of well-defined rules and mechanisms.

Even if a widely accepted system of contingency rules defining a particular perspective does not exist for populations in general, certainly there are countless examples for populations in particular. Much of natural resources management is about population management. This is particularly obvious in the more applied fields (e.g. rangeland, fisheries, forestry, and pest management, as well as epidemiological ecology), all of which require highly constrained applications of general theories to particular cases.

Theory and contingencies in community ecology

The origins of the community concept and its construction as an onto-logical entity are woven deeply in the history of ecology with Möbius, Warming and Clements being major intellectual stepping stones in its realization.[11] It might be argued that "community" has been the object of more philosophical debate than any other of ecology's entities. Debates might be framed in terms of ontological commitments and metaphysical properties. The classic example of philosophical debate in the community realm is the theory of community succession, a process that lies at the heart of both basic and applied ecology. Succession has long been one of ecology's most fruitful theories and most denigrated concepts.[12]

As early as 1904, and repeatedly thereafter, Clements and his colleagues articulated a general theory on how plant population composition and concomitant environmental change occurred with the appearance of a new land surface, or with disturbance to an old one.[13] From the outset, Clements himself recognized that succession varied with initial conditions (primary, secondary, and anomalous succession) and with moisture supply (xeric, mesic, and hydric conditions). This recognition amounted to establishing bounding contingencies which allowed detailed specification and led to better predictions, albeit in a formal classification that further dissatisfied his critics. As more was learned about succession, more perspectives in the form of particular limits were developed to improve predictions in particular instances,[14] and most recently ecologists have included alternative stable states.[15]

We acknowledge that there are other aspects to this controversy and alternative ways to frame the debate, but from the perspective of this book, our view is that one of the best known philosophical debates in ecology, ranking with the debate over density-dependent versus density-independent population regulation, was between dueling philosophical perspectives in community dynamics. Clements extended his allegedly fixed views of succession, and Gleason advanced his more stochastic observations about ecological change.[16] We could argue that this entire debate was about profound differences in ecological ontology and metaphysics, and that the antagonists would have found much agreement if they had accepted that there were contexts in which either a realist or a nominalist perspective was more appropriate (or useful), but we will not delve further in this well-plowed ground.

A later incarnation of the sense of succession and a characteristic endstate was the concept of stability, a different way of conceptualizing change and stasis. This was a more palatable view of much the same thing in a time when systems science and its associated mathematical modeling came into vogue.[17] McCoy and Shrader-Frechette analyzed the history of ecology's struggle to define stability in a meaningful way. While these authors attribute stability to a community ecology effort, it was heartily endorsed by ecosystem scientists at the same time. McCoy and Shrader-Frechette concluded that, "The most recent problems with the stability concept appear to indicate that it may be approaching heuristic bankruptcy, and that its problems may not be ones that are likely to lead us to greater conceptual clarification in community ecology."[18]

Although the debate about succession at the holistic level is one of our iconic ecological conflicts, we need to recognize that, in fact, virtually all community ecologists refer to a limited set of species populations, closer to the meaning of guilds. As Schoener[19] suggested, ecologists focus on "similia-communities" rather than the inclusive "biotic community" envisaged by Clements and Shelford at the biome level and by others like Odum more generally.[20] Thus, the ontological commitment of mainline "community ecology" is to an intermediate level of organization between population and biotic community. In this condensed context, Cody and Diamond[21] heralded the triumph of general theory in the community sphere:

> When this era began in the 1950s, ecology (*sensu "community ecology"* [our italics]) was still mainly a descriptive science. It consisted of qualitative, situation-bound statements that had low predictive value, plus empirical facts and numbers that often seemed to defy generalization. Within two decades

new paradigms had transformed large areas of ecology into a structured, predictive science that combined powerful quantitative theories with the recognition of widespread patterns of nature. This revolution in ecology had been due largely to the work of Robert MacArthur.

Even in this celebratory volume, however, there were those who recognized that overly generalized models might be less predictive than touted, and that "situation-bound statements" were still required. For example, Connell[22] was dubious that a community's species composition could be predicted exclusively with competitive theory and presented a "model" incorporating time-varying environmental and life history factors in a diagram of conditional constraints. This diagram is a perfect example of a system of contingency rules for predicting community composition.

Interestingly, an edited volume published 11 years later by Diamond and Case had a much more modest assessment of the triumph of idealization in community ecology. By self-admission, this volume was more concerned with differences between communities than with the similarities.[23] Wiens *et al.* for example, introduced a series of chapters on the importance of perspective via scale in all aspects of community-level investigations.[24] At last, spatial and temporal scales were shown to be the *sine qua non* of boundary conditions necessary to delimit theoretical statements.

Schoener's chapter in this volume is of particular importance with respect to formally elaborating rationalist theory with contingencies.[25] Schoener acknowledged that although MacArthur was, in many ways, the champion of rationalist theory in community ecology, even he realized that these were insufficient by themselves. Schoener paraphrased MacArthur's 1972 statement as:

> I predict that there will be erected a two- or three-way classification of
> organisms and their geometrical and temporal environments, this classification
> consuming most of the creative energy of ecologists. The future principles
> of the ecology of coexistence will then be of the form "for organisms of type A,
> in environments of structure B, such and such relations will hold."

Thus, MacArthur himself realized the requirement for perspective to create limiting contingencies. Stimulated by MacArthur's prediction, Schoener then presented an interesting set of axes for the organization of predictive models – essentially a framework of variables that may represent the best attempt to establish contingency rules for any aspect of ecology, in this case, for predicting community composition. Referring to Schoener's "heroic" attempts to define contingency operandi, Lawton wrote:[26]

> That nobody has seriously attempted to use and test Schoener's model in the
> decade since it was published suggests that it may all be too complicated and too

difficult to be useful ... In sum, community ecology may have the worst of all worlds. It is more complicated than population dynamics, so contingent theory does not work, or rather, the contingency is itself too complicated to be useful.

In spite of Lawton's pessimistic conclusion, selected areas of community ecology continue to seek methods for characterizing community properties, such as the continuing effort to explain and predict animal guild composition in time and space using assembly rules[27] which are termed "coalescences" in other contexts.[28] However, rather than constructing contingencies of the sort MacArthur and Schoener described, most of these rules are variations of rationalist theory related to island biogeography. The pursuit of assembly rules remains controversial and unresolved[29] and might be viewed as an effort to create a set of contingency-bound conditions for the application of rationalist theory.[30]

While Lawton despaired of ever developing useful predictive theory for communities at a species composition level of resolution, he was quite optimistic about what he and others (perhaps originally, Brown and Maurer) term "macroecology" – scale independent statistical patterns in the types, distributions, abundances, and richness of species.[31] Lawton's examples of successful macroecology include: canonical log normal distribution of population abundances, the theory of island biogeography, species-area relationships, local species abundance and geographical range relationships, and energy flow-species richness relationships. To these we would add the metabolic theory for organismal size (Brown and West 2000).[32] In general, macroecology reveals important ecological relationships – and thus theory – by the statistical comparison of many cases over orders of magnitude of spatial scale. The discerned relationships may be remarkably regular over large scalar ranges, but variation within small segments of a range reveals that local resolution may be too coarse to satisfy prediction for particular cases.[33] For such needs, other factors – perhaps contingency limits to the macroecological theory – may allow useful prediction. This illustrates again, that whether the resolution of a theory's prediction is satisfactory or not depends on the pragmatic, perspectival goals or needs of the investigator.

As a final example that pragmatic pluralism is alive and well in some areas of community ecology we can look at the effort to define and quantify interaction strength within multi-species systems. Network analysis and graph theory are inviting approaches for developing hypotheses and identifying internal structures in complex networks, but these methods require rather pragmatic ontological commitments about the nature of interactions. And perhaps the most explicit recognition of the importance of interest-based perspectivism is provided by Wootton and Emmerson:[33] "What is interaction strength? Although the concept seems intuitive, recent reviews highlight the wide diversity of definitions ... which are shaped by the interests and goals of individual investigators and, to some extent, by the empirical data available."

Theory and practice in ecosystem ecology

The ecosystem concept, or ontology, is a perspective in itself, while overlapping with what ecologists recognize as the landscape and biosphere. Even within the ecosystem perspective, however, there are multiple ways of focusing on this entity (Box 10.2). The classic approach has been through the lens of energy flow as first promulgated by Lindeman and so ably promoted in different ways by both E. P. Odum and H. T. Odum.[34] However, since the late 1960s, the material flow lens has become at least as familiar (Box 10.2). Reiners pointed out the obvious linkage of material transport and cycling with the energy-based perspective.[35] That suggestion was later richly elaborated and advanced by Sterner and Elser[36] to the extent that a subfield termed "ecological stoichiometry" was born. In fact, ecosystems might well be examined by yet a third set of principles based on information embodied in the ecosystem genome, learned signaling and reception by animals, and in physiological responses of organisms to stimuli as explicated by Reiners and Driese.[37] Ecosystems are often perceived as collective compartments connected by flows of energy and matter,[38] but they may also be perceived as networks of participating individuals and species,[39] or collectively as a dynamic system operating across fields of complex behavior with time and perturbations (Box 10.2).[40] These latter constructions are two different kinds of topological manifestations of ecosystems.[41]

As with community ecology, ecosystem ecology has a mixed history of general (realist) theory together with a large body of highly specified predictive tools often based on the concatenation of theories. The former provides predictive power of low resolution analogous to, or perhaps, as part of macroecology. Lawton noted this:[42] "In what is typically regarded as the heartland of ecosystem science, there are also big bold patterns galore. Although they have been developed by a quite different group of scientists, for quite different reasons, they have all the hallmarks of macroecology, not least a scant regard for the myriad of detailed species interactions within them."

Box 10.2
Multiple perspectives of ecosystems

Ecosystems can be conceptualized through at least four perspectives because the ecosystem ontology is loaded with a rich array of metaphysical properties. Energetic properties are illustrated in the familiar first diagram with boxes representing trophic levels, arrows representing directions of flow and widths of channels indicating the amount of energy passing through the serial array of trophic levels (redrawn from Odum and Pinkerton).[43] The second diagram illustrates material storage and flow into, out of, and within the ecosystem (authors' diagram). The third diagram illustrates a network of relationships between species populations, or possibly guilds or feeding groups as a topological network.[43] The fourth diagram illustrates dynamic behavior of ecosystems with respect to a disturbance like fire or insect pest outbreak.[44]

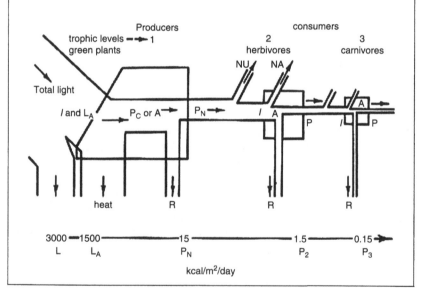

Box 10.2 (cont.)

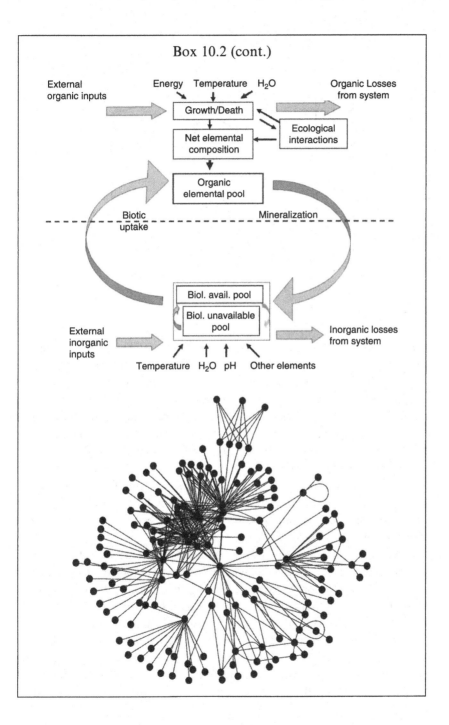

Box 10.2 (cont.)

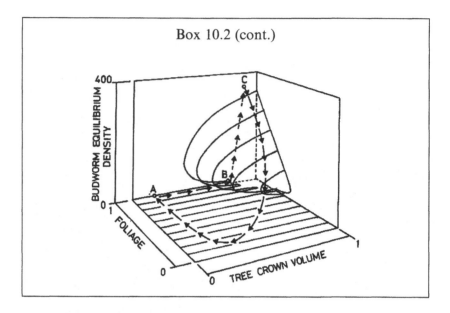

Examples of these might include: the power-efficiency hypothesis,[43] general theory of trophic level controls,[44] resource/energetics-maturity relationships,[45] successional traits of ecosystems,[46] trophic cascades,[47] adaptive cycles,[48] and ecological stoichiometry.[49] As with community-based macroecology, these are very useful theories for understanding nature in a generalized sense, and for first-order predictions of low resolution. But, these theories are also perfect examples of the dilemma expressed as Zadeh's Law of Incompatibility:[50] "as the complexity of a system increases, our ability to make precise yet significant statements about its behavior diminishes until a threshold is reached beyond which precision and significance (or relevance) become almost mutually exclusive characteristics."

Even a superficial review of mainline ecological journals indicates that the vast majority of ecosystem science favors nominalism over realism, however. That is, most ecosystem research is directed at understanding particular, even individual, systems. This may be motivated by practical necessities of society and resultant funding availability, or by a belief that by understanding the behavior of many individual systems ecologists will eventually formulate general rules for all, or at least broad classes of, ecosystems. This latter aspiration might be realized if efforts were actually made to synthesize the patterns of behavior through some kind of meta-analyses, but this is rarely done. At best, the application of general theory to specific cases with local constraints and modifications is a piecemeal practice of deploying limiting conditions to generalizations that might be termed bounded theories. Inasmuch as this is done as needed to address particular kinds of problems, it illustrates the pragmatic qualities of ecosystem ecology.

How we perambulate through our research careers

We cannot claim to know how other ecologists actually make decisions about what they choose to do for research, much less how they decide to do it. We suspect that few ecologists really have much self-awareness of these decisions based on our own experience and from questioning others (see also Box 10.1). Although it would be audacious of us to state how others think, we suspect that few ecologists set out to attack a fundamental theoretical problem, but more often describe a phenomenon in a specific system (e.g. yet another litter-fall measurement in yet another forest), and follow-up with some speculation. It is possible that many scientists subconsciously select a conceptual scale that accords with their personal potential to make a substantive change in their field (counting butterfly visitations to a flowering species to add another brick to the pollination ecology structure). Possibly, many scientists choose to work at a sufficiently small conceptual scale so that within their highly localized/specialized context, they can, in fact, have a reasonable opportunity to make a substantive change in their "field." That said, we have to agree with Peters when he wrote: "Individual scientists might dream of making basic changes in their field through fundamental new insights but realistically, most make their contributions at the level of the average and the regression."[51]

It might be interesting to interview ecologists in a non-threatening setting and talk with them about why they pursued what they did. To what extent did the lenses and skills they developed derive from hunting, fishing, collecting, or bird-watching experiences? To what extent were they attracted to do what they did because of the beauty and enticement of organisms, places, and systems? How much was a sense of public service, or service to the global environment a driving force? In this spirit, we offer a few examples of our own personal accounts through our research careers. In these one can see that predispositions by interest and training together in the context of opportunity led us through our career paths.

How wants, needs and chance drive science: tales from Reiners' experience

Earlier in my career, while at Dartmouth College, I was interested in the ecosystem aspects of terrestrial succession, particularly those involving nutrient cycling processes. I also had an interest in how ecosystem processes varied over the landscape of New Hampshire's White Mountains. Coming from the flat fields and industrial cities of the Midwest, I had a romantic fascination with the mountains and unbroken forests of northern New England as well as the physical challenges of hiking through them in all kinds of weather. Thus, youthful aesthetic inclinations played a role in setting my interests. Dartmouth was also historically involved in the origins of the Hubbard Brook project – a lively center for biogeochemical research.[52] Thus, historical excitement and opportunity played a role in my pursuits.

In 1971, Peter Vitousek was attracted to these activities and began collaborating as graduate student – an extraordinarily fortunate event for me. The confluence of location, the blossoming of the modern environmental movement, hot issues of acid rain and clear-cut forestry, awareness of the Hubbard Brook findings, Vitousek's extensive data collections in the White Mountains, and the mutual stimulation between us led to insights on the balance of nutrient inputs and outputs of terrestrial ecosystems. Building on the straightforward logic of the conservation of matter, we deduced a general theory that the inputs and outputs through an ecosystem would be controlled by changes in storage within the ecosystem.[53] This was a rationalist development of a general theory stimulated by our own data and data from Hubbard Brook. In brief, we inferred that in ecosystems, such as forests, where plant biomass was large compared with inputs, the outputs would be a function of changes in biomass. Further, the influence of biomass storage with respect to inputs and outputs would vary by element, depending on that element's stoichiometric contribution to accumulated biomass compared with its inputs and outputs. Vitousek and I realized that such a theory was at odds with the currently held idea that with increasing maturity, ecosystems would become more retentive of biogenic elements.[54] Thus, we presented our theory as an alternative to the conventional belief of the time. Thirteen years later, Haines-Young and Petch [55] analyzed this theoretical evolution as a clash of theories, highlighting it as an illustration "that theory development does not occur spontaneously ... theory arises in the context of a complex framework of ideas which includes other scientific theories."

Box 10.3
The foundations for contingent diagnoses of N-loss

Expanded description of possible limiting constraints on nitrification and transport of nitrate ions in forest ecosystems is outlined in the figure below.

A. Nitrogen mineralization/ammonification:
1. ammonium produced is immobilized by microbes in a high C/N soil environment, or
2. ammonium produced becomes fixed on and in clays, or
3. ammonium is converted to ammonia in high pH soils and lost by volatilization, or
4. ammonification is low due to chemical inhibitors or winter-cold/summer-dry conditions.

B. Conversion of ammonium to nitrate (nitrification):
1. the environment imposes an extended lag on nitrification due to chemical inhibitors, or
2. nitrate produced in one part of the soil is reduced back again to ammonium as an electron receptor, or
3. nitrate produced is reduced to volatile gases (NO, N_2O, N_2) in spatially or temporally localized microsites and lost to the atmosphere, or
4. nitrate is immobilized by microbes in a spatial is temporally differentiated part of the soil.

C. Nitrate transport through the soil profile:
1. nitrate becomes immobilized on positively charged clays below the site of production, or
2. soils have insufficient water flux through the profile to transport nitrate, or
3. partial displacement flow (percolating water follows large flow paths preventing chemical interaction with the bulk of the soil), or
4. although transported through the profile, nitrate becomes reduced at depth where redox is low (same as B3 but lower in profile).

From Vitousek *et al.* 1982.

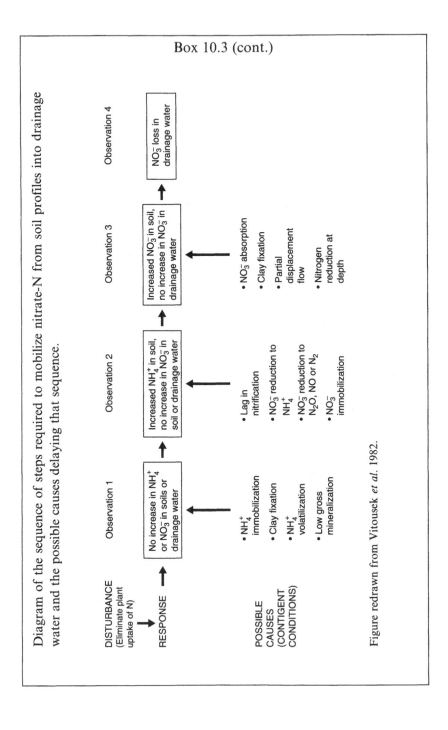

Box 10.3 (cont.)

Diagram of the sequence of steps required to mobilize nitrate-N from soil profiles into drainage water and the possible causes delaying that sequence.

Figure redrawn from Vitousek *et al.* 1982.

This story goes on. Earlier in my career – through another piece of remarkable fortune – I had befriended the noted wetlands ecologist and biogeochemist, Eville Gorham. Upon reading our paper, Gorham, in his critical and analytical way, identified numerous exceptions to our general theory. Gorham collegially communicated with us, soon leading to a joint paper describing how contingencies for classes of systems (e.g. primary versus secondary succession) would vary outcomes under the mass balance theory.[56] Thus, a broader range of empirical observations led to contingency rules for the general theory, making the theory far more predictive and informative to ecosystem ecologists. This story also illustrates how collegial interactions, brought about by chance, can direct the paths of scientific careers. Science history is full of these stories.

This perambulation continues. One of the first watershed-scale experiments carried out at Hubbard Brook was the clear-cutting and prolonged herbicide treatment of a watershed done to test effects on runoff. But as it turned out, this hydrological experiment exposed a dramatic biogeochemical reaction.[57] Shortly after cutting, the watershed's drainage waters showed very high levels of nitrate-N where it had been scarcely measurable before. This indicated that a limiting nutrient was being lost from the watershed through the clear-cutting disturbance. In fact, nitrate levels were so high as to exceed public water supply standards so a pollution issue arose as well. Besides these biogeochemical and health considerations, clear-cut logging as a silvicultural practice was just then being (re-)introduced to this part of New England and elsewhere. Some citizens objected to this practice on ecological grounds (loss of habitat, etc.), but possibly their actions were driven as much by aesthetic objections as by ecological ones. As a result, clear-cut management became an environmental cause célèbre at the time in New England and throughout the nation. Opponents seized upon the nitrate-N losses from Hubbard Brook as a scientific reason to stop the practice. The problem was, the same kinds of losses were not registered in other clear-cutting sites across the nation.

Incidental to this issue, Vitousek and I attended an open-ended workshop on biogeochemistry in 1977 at the historic Stanley Hotel[58] in Estes Park, CO. In a break-out group, a subset of us came to discuss the dilemma of why nitrate-N was lost from some clear-cut operations and not others. To my knowledge, there was no preconception by any of us that this would be a topic at this workshop. This discussion became a model of a creative, scientific event in which knowledgeable individuals interacting in a communal spirit under the right facilitator (Peter Vitousek) were able to address a phenomenon of fundamental and applied importance in an ideal, scientific manner. This group reviewed its collective understanding of the nitrogen cycle, brain-stormed factors that might prevent expected outcomes, and outlined a set of experiments to eliminate alternative inhibiting phenomena that could explain the differences observed across the country. This discussion led to a National Science Foundation-funded research award to a team of collaborators from six areas of the nation. Members conducted a sequence of uniformly applied field and laboratory experiments over a wide range of forest types to reveal the conditions required for the theoretical process to go to completion. These included about a dozen necessary proximal conditions related to soil, plant chemistry, and climatic variables arranged in a phenomenological sequence (Box 10.3). This team sought the critical conditions or constraints that prevented completion of the process from removal of plants to fluvial loss of nitrate-N. Conversely, the team established the contingencies necessary for the process to go to completion.

The results of this team research are exhibited in Box 10.3. Except for cases of gaseous loss (A3, B3 and C4) these constraints are just delaying mechanisms for nitrate-N loss, although delays may last several years, to the time when plant uptake is re-established. If plant growth is not resumed before the delay is saturated, nitrate losses will inevitably occur. With the identification of these conditions, Vitousek *et al.*[59] established contingencies that had to be associated with a general theory in order to better understand how nature works, and to predict the effects of this kind of perturbation. The formalizations shown in Box 10.3 stop just short of creating a diagnostic tool for predicting nitrogen loss due to clear-cutting by applying a system of contingencies.

How wants, needs and chance drive science: tales from Lockwood's experience

Shortly after coming to the University of Wyoming, I began to work on the biological control of rangeland grasshoppers with *Nosema locustae*.[60] This research reflected my values with respect to minimizing harm to nontarget organisms, which were often decimated by broad-spectrum insecticides. The pragmatic concern was that the pathogen was not providing adequate control of pest populations to justify its short-term economic costs. This led to a series of investigations focused on understanding the limiting step in the epizootic process – the transmission of the pathogen through necrophagy of diseased individuals by healthy grasshoppers.

From these investigations, it became evident that grasshoppers rapidly located cadavers via chemical cues, and through some clever work by my graduate student, Chuck Bomar, it became clear that oleic, linolenic, and linoleic acids (via the breakdown of body lipids) were the attractant and phagostimulant cues.[61] Not long after this discovery, a dramatic change in the funding of grasshopper control programs took my research in a different direction which would eventually return to the role of semiochemicals.

The federal government withdrew its generous subsidy for rangeland grasshopper control on private, state, and federal lands, effectively doubling or tripling the cost to the rancher. By good fortune, I had recently returned from a sabbatical leave in Australia, where I'd spent some time with the Plague Locust Commission. This pest management organization was using an insecticide application method in which swaths of chemical were applied at intervals, rather than as a continuous blanket. The fast-moving locusts might safely pass through an untreated area but they'd inevitably come into contact with a treated swath.

My research associate, Scott Schell, and I reasoned that although rangeland grasshoppers were not as mobile as locusts, perhaps they sufficed as a natural kind with regard to management. If so, then insecticide could be applied at intervals and still provide acceptable control on US rangelands. Our initial studies showed that applying carbaryl to just every other swath of rangeland provided nearly 90% mortality, which compared very favorably to the 95–98% control with blanket applications. By reducing the rate of application as well as the coverage, we were able to provide ranchers with a method that reduced their costs by three-fourths with only a 10% decrease in control.[62]

The reason why high rates and complete coverages had been the standard approach was that the USDA (the agency conducting the control programs and setting the rules by virtue of providing a subsidy) wanted to assure public (rancher/landowner) satisfaction. From the agency's perspective, the economic costs of over-application were offset by the political costs of a failed program. As such, their interests dictated the nature of the ecological intervention.

With the shift of the economic burden to the private beneficiaries, the interest became one of return on investment. In terms of this pragmatic standard, our approach was eminently successful. Indeed, it is now the standard method used across the western United States and has saved

millions of dollars and thousands of pounds of insecticide. But we didn't
know the mechanism by which it worked – movement of grasshoppers
into the treated areas seemed plausible, but the conservation of natural
enemies that would otherwise have been eliminated with a blanket application could also have accounted for at least some of the success.[63]

Driven by an interest in improving the method, I worked with another
graduate student, Narisu, to ascertain the directionality and movement
of rangeland grasshoppers. We discovered, among other factors, that
these insects tended to move into the prevailing wind which meant
dispersal was typically east-west.[64] This pattern accorded ideally with
the normal orientation of the insecticide swaths, which were north-south
(meaning that the grasshoppers had a high probability of moving into a
treated area) so that the pilot avoiding flying into the rising sun (most
applications occurring in the early morning). We also found that
the necrophilic grasshoppers were attracted to the dead insects in the
insecticidal swaths.

Given these ecological considerations, we returned to the matter of
fatty acids as attractants. Schell found that some vegetable oils were very
high in oleic, linolenic, and linoleic acids. These oils functioned to stimulate both movement and feeding in laboratory and small-plot field trials.
We then tested formulations of insecticides with canola oil as a carrier
and found that 85–90% control could be maintained while increasing the
distance between treated swaths and decreasing the rate of insecticide
within swaths.[65] So, the moral value of minimizing harm to nontarget
organisms and the economic value of reducing costs to producers were
simultaneously addressed.

Although we had a generically efficient method of control, local
economic and ecological considerations as well as large-scale allocations
of resources (e.g. state-funded surveys) were yet to be addressed.
To understand population dynamics at the state and regional level,
I collaborated with my brother, Dale Lockwood (who was trained in both
applied mathematics and ecology), to develop predictive models using the
principles of catastrophe theory and self-organized criticality. Underpinned
by deductive, rationalist considerations, these mathematical approaches
were empirically tested against observed dynamics – and the results
strongly suggested that complex, nonlinear phenomena were central to
the population ecology of these insects.[66] Along with these studies, Schell
and I used GIS (geographic information systems) to analyze the spatial
patterns and ecological associations of grasshopper outbreaks and
found that soil was the best predictor of frequently infested lands.[67]

This information allowed a far more efficient partitioning of the state's surveyors, with the intensity of the survey being matched to both static and dynamic features (e.g. Markov chain analysis was used to estimate the likelihood of a population increasing from innocuous to economically damaging levels).

In an effort to convert ecological knowledge and management experience into a form accessible to pest managers, I collaborated with Karl Branting, a faculty member in computer science at the University of Wyoming, and his graduate student, John Hastings. Our work led to the development of CARMA (Case-based Rangeland Management Advisor).[68] This software included elements of case-based, rule-based, and probabilistic reasoning to generate recommendations indistinguishable from those of a panel of human experts in response to a particular infestation, the parameters of which were provided by a rancher with little or no ecological or entomological expertise. The CARMA platform was well-received by practitioners and is now being extended to other western states. This operational decision-support tool does not look like conventional theory, but embodies general principles driven from the perspective of local contingencies.

It may be just a short step between decision-support tools for managing extensive ecosystems, to generating diagnostic tools like those used in medical practice for determining outcomes of particular situations. In fact, the similarities between the practice of medicine and applied ecology have been made before. Such diagnostic systems would be an advanced means of attaching limiting conditions to ecological theories.

Endnotes

1. Schoener, T. W. 1986a. Mechanistic approaches to ecology: a new reductionism? *American Zoologist*, **26**: 81–106; Schoener, T. W. 1986b. Overview. Kinds of ecological communities: ecology becomes pluralistic. In *Community Ecology*, eds. J. Diamond and T. J. Case. New York: Harper and Row, 467–479; Allen, T. F. H. and T. W. Hoekstra. 1992. *Toward a Unified Ecology*. New York: Columbia University Press; Ulanowicz, R. E. 1999. Life after Newton: an ecological metaphysic. *BioSystems*, **50**: 127–142.
2. Instrumentalism might be taken as making a stronger claim than we wish to assert. A hard-driving instrumentalist might argue that science does not aim at truth, wherein truth is taken to mean objective reality. We'd agree that science does not succeed in mirroring reality in a correspondent manner, but constrained perspectivism entails that when we act on our beliefs the actual world can "push back" such that we can know something about reality through empirical feedback. As such, science can probe the limits of objective existence, although it is perhaps most common that we limit our concepts via social,

cultural, or other nonphysical constraints. But even these limits are a kind of objective quality of the world in a pan-realistic sense.

3. Pickett, S. T. A., J. Kolasa and C. G. Jones. 2007. *Ecological Understanding. The Nature of Theory, the Theory of Nature.* 2nd edn. San Diego, CA: Academic Press.
4. Schrader-Frechette, K. S. and E. D. McCoy. 1993. *Methods in Ecology: Strategies for Conservation.* Cambridge, UK: Cambridge University Press.
5. Wimsatt, W. C. 2007. *Re-engineering Philosophy for Limited Beings: Piecewise Approximations to Reality.* Cambridge, MA: Harvard University Press.
6. Gause, G. F. 1934. *The Struggle for Existence.* Baltimore: Williams and Wilkins; Wolfenbarger, D. O. 1946. Dispersion of small organisms, distance dispersion rates of bacteria, spores, seeds, pollen, and insects; incidence rates of diseases and injuries. *The American Midland Naturalist*, **35**: 1–152.
7. Huffaker, C. B. 1958. Experimental studies on predation: dispersion factors and predator-prey oscillations. *Hilgardia*, **27**: 343–383; Rhodes, O. E., Jr. and E. P. Odum. 1996. Spatiotemporal approaches in ecology and genetics: the road less traveled. In *Population Dynamics in Ecological Space and Time*, eds. O. E. Rhodes, Jr., R. K. Chesser and M. H. Smith. Chicago: The University of Chicago Press, 1–7.
8. Lotka, A. J. 1920. Analytic note on certain rhythmic relations in organic systems. *Proceedings of the National Academy of Sciences*, **6**: 410–415; Lotka, A. J. 1956. *Elements of Mathematical Biology.* Mineola, NY: Dover.
9. Morris, R. F., ed. 1963. The dynamics of epidemic spruce budworm populations. *Memoirs of the Entomological Society of Canada* 31. Ottawa: Entomological Society of Canada.
10. Lawton, J. H. 1999. Are there general laws in ecology? *Oikos*, **84**: 177–192, 179.
11. McIntosh, R. B. 1985. *The Background of Ecology: Concept and Theory.* Cambridge, UK: Cambridge University Press, 209.
12. Keller, D. R. and F. B. Golley. 2000. *The Philosophy of Ecology: From Science to Synthesis.* Athens, GA: University of Georgia Press.
13. Clements, F. E. 1916. *Plant Succession: An Analysis of the Development of Vegetation.* Washington, D.C.: Carnegie Institution of Washington; Clements, F. E. 1905. *Research Methods in Ecology.* Lincoln, NB: The University Publishing Company; Weaver, J. E. and F. E. Clements. 1929. *Plant Ecology.* New York: McGraw-Hill Book Co.; Clements, F. E. and V. E. Shelford. 1939. *Bio-ecology.* New York: John Wiley & Sons.
14. Egler, F. E. 1954. Vegetation science concepts I: initial floristics composition, a factor in old-field vegetation development. *Vegetatio*, **4**: 412–417; Peet, R. K. 1981. Changes in biomass and production during secondary forest succession. In *Forest Succession: Concepts and Application.* Springer-Verlag; Glenn-Lewin, D. C., R. K. Peet, T. T. Veblen, eds. 1992. *Plant Succession: Theory and Prediction.* New York: Chapman & Hall.
15. Tongway, D. and J. Ludwig. 2005. Heterogeneity in arid and semi-arid lands. In *Ecosystem Function in Heterogeneous Landscapes*, ed. G. M. Lovett, C. Jones, M. G. Turner and K. C. Weathers. New York: Springer, pp. 189–205; Schröder, A., L. Persson and A. De Roos. 2005. Direct experimental evidence for alternative stable states: a review. *Oikos*, **110**: 3–19; Didham, R. K., C. H. Watts and D. A. Norton. 2005. Are systems with strong underlying abiotic regimes more likely to exhibit alternative stable states? *Oikos*, **110**: 409–416.
16. Keller and Golley, *The Philosophy of Ecology.*

17. McIntosh, R. P. 1985. *The Background of Ecology: Concept and Theory.* Cambridge, UK: Cambridge University Press, 209.
18. McCoy, E. D. and K. Schrader-Frechette. 1992. Community ecology, scale, and the instability of the stability concept. *Philosophy of Science*, 1: 184–199, 193.
19. Schoener, Overview. Kinds of ecological communities, 469.
20. Clements and Shelford, *Bio-ecology*; Odum, E. P. 1971. *Fundamentals of Ecology.* 3rd edn. Philadelphia: W. B. Saunders Company.
21. Cody, M. L. and J. M. Diamond. 1975. Preface. In *Ecology and Evolution of Communities*, eds. M. L. Cody and J. M. Diamond. Cambridge, MA: Belknap Press of Harvard University Press, vii–ix, vii.
22. Connell, J. H. 1975. Some mechanisms producing structure in natural communities: a model and evidence from field experiments, In *Ecology and Evolution of Communities*, eds. M. L. Cody and J. M. Diamond. Cambridge, MA: Belknap Press of Harvard University, 460–489.
23. Diamond, J. M. and T. J. Case, eds. 1986. *Community Ecology.* New York: Harper and Row.
24. Wiens, J. A., J. F. Addicott, T. J. Case and J. M. Diamond. 1986. Overview: the importance of spatial and temporal scale in ecological investigations. In *Community Ecology*, eds. J. M. Diamond and T. J. Case. New York: Harper and Row, 145–153; Wiens, J. A. 1999. Toward a unified landscape ecology. In *Issues in Landscape Ecology*, eds. J. A. Wiens and M. R. Moss, Ft. Collins, CO: International Association for Landscape Ecology, 148–151.
25. Schoener, Overview. Kinds of ecological communities, 486.
26. Lawton, Are there general laws in ecology?, 182.
27. Diamond, J. M. 1975. Assembly of species communities. In *Ecology and Evolution of Communities*, ed. M. L. Cody and J. M. Diamond. Cambridge, MA: Belknap Press of Harvard University Press, 342–444; Morton, R. D., R. Law, S. L. Pimm and J. A. Drake. 1996. On models for assembling ecological communities. *Oikos*, **75**: 493–499; Belyea, L. R. and J. Lancaster. 1999. Assembly rules within a contingent ecology. *Oikos*, **86**: 402–416; Ulrich, W. 2004. Species co-occurrences and neutral models: reassessing J. M. Diamond's assembly rules. *Oikos*, **107**: 603–609; Morris, D. W. 2005. On the roles of time, space and habitat in a boreal small mammal assemblage: predictably stochastic assembly. *Oikos*, **109**: 223–238.
28. Morin P. J. 1999. *Community Ecology.* Malden, MA: Blackwell Science; Thompson, J. N., O. J. Reichman, P. J. Morin, *et al.* 2001. Frontiers of ecology. *BioScience*, **51**: 15–25.
29. e.g. Morin, *Community Ecology*; Thompson *et al.* Frontiers of ecology.
30. Temperton, V. M., R. J. Hobbs, T. Nuttle and S. Halle, eds. 2004. *Assembly Rules and Restoration Ecology: Bridging the Gap Between Theory and Practice.* Washington, D.C.: Island Press.
31. Brown, J. H. and B. A. Maurer. 1989. Macroecology: the division of food and space among species on continents. *Science*, **243**: 1145–1150.
32. Brown, J. H. and G. B. West. 2000. *Scaling in Biology.* New York: Oxford University Press.
33. Wootton, J. T. and M. Emmerson. 2005. Measurements of interaction strength in nature. *Annual Review of Ecology, Evolution, and Systematics*, **36**: 419–444.
34. Lindeman, R. L. 1942. The trophic-dynamic aspect of ecology. *Ecology*, **23**: 399–418; Odum, E. P. 1953. *Fundamentals of Ecology.* Philadelphia:

W.B. Saunders Co.; Odum, H. T. 1971. *Environment, Power and Society.* New York: Wiley-Interscience.

35. Reiners, W. A. 1986. Complementary models for ecosystems. *The American Naturalist*, **127**: 399–418.
36. Sterner, R. W. and J. J. Elser. 2002. *Ecological Stoichiometry: The Biology of Elements from Molecules to the Biosphere.* Princeton, NJ: Princeton University Press.
37. Reiners, W. A. and K. L. Driese. 2004. *Transport Processes in Nature: The Propagation of Ecological Influences Through Environmental Space.* Cambridge, UK: Cambridge University Press.
38. Odum, *Environment, Power and Society.*
39. Montoya, J. M. and R. V. Solé. 2003. Topological properties of food webs: from real data to community assembly models. *Oikos*, **102**: 614–622.
40. Gunderson, L. H. and C. S. Holling, eds. 2002. *Panarchy: Understanding Transformations in Human and Natural Systems.* Washington, D.C.: Island Press; Roopnarine, P.D. 2007. Ecological complexity: Catastrophe Theory. In *Encyclopedia of Ecology,* ed. S. E. Jorgensen. Oxford: Elsevier.
41. Prager, S. D. and W. A. Reiners. 2009. Historical and emerging practices in ecological topology. *Ecological Complexity*: doi: 10.1016/j.ecocom.2008.11. 001; Odum, E. P. 1963. *Ecology.* New York: Holt, Rinehart and Winston, 38; Montoya, J. M. and R. V. Solé, Topological properties of food webs; Holling, C. S. 1981. Forest insects, forest fires, and resilience. In *Fire Regimes and Ecosystem Properties: Proceedings of a Conference,* eds. H. A. Mooney, T. M. Bonnicksen, N. L. Christensen, J. E. Lotan and W. A. Reiners. US Forest Service General Technical Report WO-26, 445–464, 452.
42. Lawton, Are there general laws in ecology?, 188.
43. Odum, H. T. and R. C. Pinkerton. 1955. Time's speed regulator: the optimum efficiency for maximum power output in physical and biological systems. *American Scientist*, **43**: 331–343.
44. Hairston, N. G., F. E. Smith and L. B. Slobodkin. 1960. Community structure, population control, and competition. *The American Naturalist*, **94**: 421–425.
45. Margalef, R. 1963. On certain unifying principles in ecology. *The American Naturalist*, **97**: 357–374.
46. Odum E. P. 1969. The strategy of ecosystem development. *Science*, **164**: 262–270.
47. Pace, M. L., J. J. Cole and S. R. Carpenter. 1999. Trophic cascades revealed in diverse ecosystems. *Trends in Ecology and Evolution*, **14**: 483–488.
48. Holling, C. S. and L. H. Gunderson. 2002. Resilience and adaptive cycles. In *Panarchy: Understanding Transformations in Human and Natural Systems*, eds. L. H. Gunderson and C. S. Holling. Washington, D.C.: Island Press, 25–62.
49. Sterner and Elser, *Ecological Stoichiometry.*
50. Zadeh L. A. 1973. Outline of a new approach to the analysis of complex systems and decision processes. *IEEE Transactions on Systems, Man, and Cybernetics*, SMC-3: 28–44, 28.
51. Peters, R. H. 1991. *A Critique for Ecology.* New York: Cambridge University Press, 20.
52. Likens, G. E., F. H. Bormann, R. S. Pierce, J. S. Eaton and N. M. Johnson. 1977. Biogeochemistry of a forested ecosystem. New York: Springer-Verlag, v.
53. Vitousek, P. M. and W. A. Reiners. 1975. Ecosystem succession and nutrient retention: a hypothesis. *Bioscience*, **25**: 376–381. In 1991 this paper became a citation classic which led to a more detailed explanation of how the paper

developed. See: Vitousek, P. M. and W. A. Reiners. 1991. Ecological
succession and nutrient budgets. *Current Contents: Agricultural Biology
and Environmental Sciences*, **22**(42): 10. Week's Citation classic.

54. Odum, The strategy of ecosystem development.
55. Haines-Young, R. H. and J. R. Petch. 1986. *Physical Geography: Its Nature
and Methods*. London: Harper and Row, 131.
56. Gorham, E., P. M. Vitousek and W. A. Reiners. 1979. The regulation of
chemical budgets over the course of terrestrial ecosystem succession. *Annual
Review of Ecological and Systematics*, **10**: 53–84. Both Gorham and Vitousek were
later elected to the National Academy of Sciences for their contributions to
ecology.
57. Bormann, F. H., G. E. Likens, D. W. Fisher and R. S. Pierce. 1968. Nutrient
loss accelerated by clear-cutting of a forest ecosystem. *Science*, **159**: 882–884;
Likens, G. E., F. H. Bormann, N. M. Johnson, D. W. Fisher and R. S. Pierce.
1970. Effects of forest cutting and herbicide treatment on nutrient budgets in
the Hubbard Brook watershed-ecosystem. *Ecological Monographs*, **40**: 23–47.
58. Location of the Jack Nicholson thriller, "The Shining."
59. Vitousek, P. M., J. R. Gosz, C. C. Grier, *et al.* 1979. Nitrate losses from
disturbed ecosystems. *Science*, **204**: 469–474; Vitousek, P. M., W. A. Reiners,
J. M. Melillo, C. C. Grier and J. R. Gosz. 1981. Nitrogen cycling and loss
following forest perturbation: the components of response. In *Stress Effects on
Natural Ecosystems*, eds. G. W. Barrett and R. Rosenberg. Chichester, UK:
John Wiley & Sons Ltd, 115–127; Vitousek, P. M., J. R. Gosz, C. C. Grier,
J. M. Melillo and W. A. Reiners. 1982. A comparative analysis of potential
nitrification and nitrate mobility in forest ecosystems. *Ecological Monographs*,
52: 155–177.
60. Lockwood, J. A. and L. D. DeBrey. 1990. Direct and indirect effects of
Nosema locustae (Canning) (Microsporida: Nosematidae) on rangeland
grasshoppers (Orthoptera: Acrididae). *Journal of Economic Entomology*,
83: 377–383.
61. Bomar, C. R. and J. A. Lockwood. 1994. The olfactory basis for cannibalism
in grasshoppers (Orthoptera: Acrididae): I. Laboratory assessment of
attractants. *Journal of Chemical Ecology*, **20**: 2249–2260; Bomar, C. R. and
J. A. Lockwood. 1994. The olfactory basis for cannibalism in grasshoppers
(Orthoptera: Acrididae): II. Field assessment of attractants. Ibid, 2261–2272.
62. Lockwood, J. A. and S. P. Schell. 1997. Decreasing economic and
environmental costs through reduced area and agent insecticide treatments
(RAATs) for the control of rangeland grasshoppers: Empirical results
and their implications for pest management. *Journal of Orthoptera Research*,
6: 19–32.
63. Norelius, E. E. and J. A. Lockwood. 1999. The effects of standard and
reduced agent-area insecticide treatments for rangeland grasshopper
(Orthoptera: Acrididae) control on bird densities. *Archives of Environmental
Toxicology*, **37**: 519–528.
64. Narisu, J. A. Lockwood, and S. P. Schell. 2000. Rangeland grasshopper
movement as a function of wind and topography: implications for pest
management. *Journal of Orthoptera Research*, **9**: 111–120.
65. Lockwood, J. A., Narisu, S. P. Schell and D. R. Lockwood. 2001. Canola
oil as a kairomonal attractant of rangeland grasshoppers (Orthoptera:
Acrididae): an economical liquid bait for insecticide formulation. *International
Journal of Pest Management*, **47**: 185–194.

66. Lockwood, J. A. and D. R. Lockwood. 1991. Rangeland grasshopper population dynamics: insights from catastrophe theory. *Environmental Entomology*, **20**: 970–980; Lockwood, D. R. and J. A. Lockwood. 1997. Evidence of self-organized criticality in insect populations. *Complexity*, **2**: 49–58.
67. Schell, S. P. and J. A. Lockwood. 1997. Spatial characteristics of rangeland grasshopper population dynamics in Wyoming: implications for pest management. *Environmental Entomology*, **26**: 1056–1065.
68. Branting, L. K., J. D. Hastings and J. A. Lockwood. 1997. Integrating cases and models for prediction in biological systems. *Artificial Intelligence Applications*, **11**: 29–48; Hastings, J. D., L. K. Branting and J. A. Lockwood. 1996. A multi-paradigm system for rangeland pest management. *Computers and Electronics in Agriculture*, **16**: 47–67.

11

What constrained perspectivism offers
to the teaching of ecology

We have proposed that constrained perspectivism is both a prescription for, and description of, how ecology is done. On one hand, we assert that ecologists adopt their perspectives while being largely unaware of what operational heuristics are in place, much like a nation persists without its citizens understanding their constitution. On the other hand, we believe that ecology would better progress if its "citizens" (and leaders) understood the formal structure of the philosophy underlying these heuristics. How can awareness and understanding of the perspectival nature of ecology be brought into the consciousness of its practitioners? To the extent that practitioners read this and similar sources, consciousness might be obtained. More broadly, though, philosophical awareness and adaptation must be achieved through the education and training of new ecologists. This chapter is devoted to considerations of how present philosophical positions (often a form of naïve realism) are implicitly inculcated through our teaching, and how those positions might be changed.

Ecology is difficult

Ecology is not an easy science to understand or teach. As we described earlier, ecologists use a remarkable range of methods and techniques to understand complex, inherently variable, functionally diverse, historically shaped entities and processes across a staggering range of spatial, temporal, and interactive scales (see Chapter 3). Add to this challenge the various philosophical positions of absolutism, relativism, and pluralism concerning what we claim to be real and known, and the result is a very difficult field for students to comprehend.

At the risk of caricaturing current practice, we suggest that ecology (along with most other academic fields) is taught in an authoritarian manner – truth is possessed by professors who deliver knowledge to students. And, at advanced levels, ecology is philosophically simplified by limiting inquiry and practice to studiously guarded philosophical and topical domains. Perhaps most insidiously, we protect our students (and ourselves) from uncomfortable challenges by defining philosophy as extraneous to professional success. We do this by defining "success" as generating fundable grants, publishing refereed manuscripts, and presenting respectable papers. These measures of success are intrinsic to the scientific culture that we created and the reward system we control. As such, they are self-renewing by their very nature. There are only weak outside forces to change the culture.

To the extent that this caricature has merit, ecology's cultural attitude toward philosophy creates serious problems. First, philosophically "flying blind" is not an optimal way of gaining knowledge about nature. We have highlighted some of the disputes based on ecologists' inability to understand opposing philosophical positions. Second, without the broader vision provided by better philosophical grounding, it is not clear "which way is forward" when it comes to advancing ecology. As a consequence, we fly in circles. Third, some (perhaps many) students will eventually in their education, or later in their careers, question ecology's presumptions of reality, knowledge, and value. Without answers for these students, we are likely to see them move onto other fields that accord with their intellectual curiosity. This would be flying into the ground in terms of losses to the discipline. We contend that avoiding the messiness of interactions with other disciplines – specifically ignoring our philosophical limitations through academic specialization – is not a viable solution.

Ecology truly is difficult when we engage science as a human venture worth probing deeply and understanding fully – or at least as fully as one can grasp such a staggeringly complex enterprise. The undergraduate who comes to an ecology class believing that the professor, as a scientist steeped in the field, possesses Truth (and, as we shall see, this characterization is not so far from the case in terms of cognitive development) and is prepared to distribute an account of the way the world really is, resembles a child encountering Santa Claus in the mall. We can pass out our gifts of knowledge wrapped in objectivity but someday the student is likely to figure out (or be told) that there's no such thing as a scientific Santa who delivers absolute truths. With this realization, the student might well feel betrayed. But a parent who anticipated the child's suspicions regarding the *Truth of Santa* might begin to lay the groundwork for an understanding of mythic figures as normative ideals. In a parallel manner, here's what we might tell our ecology student (let's call her Virginia) who has come to doubt the *Truth of Science* because her classmates taking a course in postmodern literary criticism have claimed that there is no truth at all:[1]

> [Your friends] are wrong. They have been affected by the skepticism of a skeptical age. They do not believe except [what] they see. They think that nothing can be which is not comprehensible ... All minds, Virginia, whether they be [adults' or students'], are little. In this great universe of ours man is a mere insect, an ant, in his intellect, as compared with the boundless world about him, as measured by the intelligence capable of grasping the whole of truth and knowledge.

A mentor might go on to assure her that, yes, there is scientific truth, and it exists as certainly as the relief of human suffering, the restoration of natural beauty, the alleviation of hunger, the conservation of species, and any of the other results of science that have empirically served the authentic interests of humanity. This is not truth in an absolute sense, just as there is no Santa Claus in a literal sense. But both truth and generosity are real insofar as believing in these has proven useful to our leading meaningful lives.

We contend that amelioration of the fruitless rifts between ecologists, cynical skepticism among students of ecology, and dimness of vision for progress, can be addressed via constrained perspectivism. Indeed, we believe that direct engagement with constrained perspectivism in the education of ecology students is central to making such change a reality. How might this be done?

Framing the educational challenges: an overview

The education of ecologists can be divided into four phases. The period of kindergarten through high school is the formative period in which environmental values are developed. This axiology is initially established through inculcation of right actions which are endorsed or enforced by authority figures at home, church, and school – and later, at least ideally, through conscious evaluation and choice. However, our experiences give us little foundation for suggesting if or how epistemological and ontological/metaphysical elements of ecology should be incorporated into this developmental phase. The second phase, baccalaureate education, is typically the first encounter with the science of ecology, at least in its quantitative formulations. Along with this comes an introduction to science as a way of knowing that is different from parental authority, peer opinion, and religious pronouncements. The third phase for a professional ecologist is graduate education, the time of professionalization in preparation for a life of work in ecology. Finally, there is the fourth and presumably life-long phase of post-degree learning in which the individual comes to understand science from the "inside," reshaped and reinforced by direct and varied experiences of the natural world and the social network of ecologists and allied professionals.

One might be tempted to suggest that this last phase has little potential in terms of education – that it is difficult to teach an old dog new tricks. To be sure, new techniques and subdisciplines can (and must) be learned, but entirely new perspectives as to the foundations of science itself seem rather less teachable. Ironically, however, we propose that professors are more likely to change their views of ecological science in the process of teaching than through the direct experience of questioning their own scholarly approach in the course of research. In fact, we believe that teaching can be a particularly effective way of reviewing old ways of framing ecological ideas and revising them in the light of new perspectives uncovered by other ecologists and scientists (Box 11.1). If one wants to truly understand an ecological concept, teach it to students who have not learned to accept the "givens" of science. In our experience, engagements with students have led to some of the most troubling questions about the implicit values of ecology. For example, conservation biology's axiological position is evident in the very label of the field, but ecologists may not recognize the implicit values and the aesthetic, ethical, or religious entailments that students perceive in the title.

Box 11.1
Experience, if not authority

The reader might well wonder by what authority we are offering our views on the teaching of ecology. While we claim no particular expertise in this regard, we can appeal to our experience as professors and hope that this assuages the reader's concern as to the foundation upon which we base our recommendations. We admit that our experiences may well be idiosyncratic, but they are, at least, reasonably extensive.

Box 11.1 (cont.)

Lockwood has taught undergraduate and graduate courses in ecology on 32 occasions to a total of more than 1000 students, with offerings that included: Biodiversity (a freshman-level general studies course), Ecology as a Discipline (the history and philosophy of ecology for doctoral students), Insect Population Biology (concepts, models and theories, focusing on outbreak and nonlinear dynamics), and The Classic Literature of Ecology (a survey of historic books and papers). Courses at the interface of ecology and philosophy have included: Agricultural Ethics, Natural Resource Ethics, Environmental Ethics and seminars in Deep Ecology, Environmental Justice, and Ecofeminism.

Reiners has taught biology, ecology, environmental studies/policy, and earth system science courses over the last 44 years at three contrasting institutions (University of Minnesota, Dartmouth College, and University of Wyoming). These have ranged from freshman level biology to graduate-level ecosystem and biogeochemistry courses. Perhaps of greatest relevance to this chapter were many years of teaching general ecology to a total of perhaps 1000 students, and nearly continuous teaching of field courses in which concepts were matched against observations in settings of natural beauty.

Together, we have taught an upper-level undergraduate course, The Origins of Environmental Beliefs, and a graduate-level course, The Philosophy of Science in Ecology. The latter served as the catalyst for our further readings and discussions, which eventually led to this book.

Baccalaureate teaching: what the students encounter

The first course in ecology

Whatever the reason undergraduates select an ecology course, and perhaps our experiences are geographically particular,[2] there can be little doubt as to the importance of the introductory class in general ecology for planting the philosophical seeds of our science. This course serves as the tuner, amplifier, and resistor of the "folk learning" that the students bring to the course through their often extensive experiences and popular readings. "Bar room biology" (or "hunting camp ecology") is a pervasive foundation for the heuristics Wyoming students bring to campus. It is often impossible to overcome this influence by the time of graduation. Courses subsequent to the first general introduction usually focus on specific aspects of ecology (e.g. community ecology, population ecology, conservation biology), where the diversity of conceptual problems is limited. The "big questions" of ecology – what is valuable, what constitutes knowledge, what is real – arise early, at least implicitly, in the introductory course. The answers are often fixed and taken to constitute the axiomatic norms in a general ecology course and these standard philosophical views become the "givens" in further studies. We now turn to the introductory textbooks, as these represent the critical source of conceptual foundations.

General ecology textbooks: the templates of our science

Why take textbooks so seriously? Regardless of instructors' contentions that textbooks are ancillary to their lectures, students regard a text as an authoritative anchor to the disciplinary standard of ecology. Presumably, these market-tested products provide a window on whatever consensus exists in ecology. Textbook language tends to be, as Stephen J. Gould noted, "nuanced and judicious."[3] In the context of synthesizing evolutionary theory, he wrote,

> To learn the unvarnished commitments of an age, one must turn to the textbooks that provide "straight stuff" for introductory students. Yes, textbooks truly oversimplify their subjects, but textbooks also present the central tenets of a field without subtlety or apology – and we can grasp thereby what each generation of neophytes first imbibes as the essence of a field. Moreover, many textbooks boast authorship by the same professionals who fill their technical writings with exceptions, caveats, and complexities. I have long felt that surveys of textbooks offer our best guide to the central convictions of any era. What single line could be more revealing, more attuned to the core commitment of a profession that bathed in the blessings of Victorian progressivism, and aspired to scientific status in Darwin's century than the epigram that Alfred Marshall placed on the title page to innumerable editions of his canonical textbook, Principles of Economics: "*natura non facit saltum*".

It is well beyond the scope of our objectives to do a complete analysis of all the possible dimensions by which one could judge textbooks, although such might be an interesting and valuable project. Textbooks

provide organizational frameworks for most, if not all, general ecology courses and in this respect they vary little from one another. But
within those frameworks there are differences in emphases (e.g. see definitions of ecology in Box 2.3). We celebrate these differences as realizations of ecological pluralism – multiple perspectives in print – but at the
same time we are interested in the philosophical implications of the
differences.

A striking commonality is that, with one exception in our survey
experience,[4] texts are organized by levels of organization or hierarchical
layers: individual, population, community, ecosystem, landscape, and
biosphere. We noted earlier that such levels are elaborations of entities
which, in turn, are foundational to the way we look at nature. Whether
made consciously or not, these are ontological commitments. One could
say that these commitments are functionally and commercially ratified by
reviewers, editors, and publishers, although the ratifications are de facto,
derived without discussion of their reality or implications.

While ontological layering is nearly universal, there are some intriguing
differences within this structure. For example, by the absence of ecosystem sections, we infer that some textbook writers do not recognize ecosystems as entities. Where ecosystem sections do exist, accounts of their
metaphysical properties are expressed in the subsections. For example,
ecosystems seem to have the qualities of species structure, energetics,
material transfer, and dynamic change in time. Interestingly, these same
properties are attributed to communities in texts not recognizing ecosystems. Processes or phenomena, as opposed to entities, are treated as
observable facts rather than as conceptual frameworks in themselves (see
ways of perceiving ecosystems in Box 6.2). Of course, metaphysical properties also are generally not explicitly justified in textbooks, although they
might be found in more focused literature. For example, boundaries are
implicit to the ecosystem ontology but are only specified for particular
examples such as the border of a field, pond, or watershed. A more
general metaphysical specification might be described as that circumfer-
ential zone in which material, energy, and information transport are
minimized (the attendant problems of becoming clear on whether material, energy, or information is the quality of interest and what constitutes
"minimized" serve to illustrate the philosophical challenges associated
with ecosystems).

Consumers of these texts – students – are typically uninterested in the
founders of the ontologies, the developers of metaphysical properties, or
the innovators of ways of knowing, and this disinterest is reflected in the

extremely limited discussion of the origins of ideas. From experience we know that the history of science can be deadly in the classroom (although much of this may be the way in which it is written and spoken about rather than its essential nature). But, by ignoring the history of ideas and attributions to real people, ontologies seem to have arisen *ex cathedra* and the adventure and contingencies of intellectual development are lost. There are some good examples, though. One can find fairly personalized and story-like descriptions of how F. H. Bormann and G. E. Likens came to realize that the watershed – an ontology in hydrology – could be viewed as revealing ecosystems through its hydrological properties, and that its biogeochemical (metaphysical) properties could become known by a budgetary system of chemical measurements.[5] Such examples can dramatize the creative leap and illustrate the philosophical thought involved in ecological realizations.

Another source of variation among texts is the treatment of rationalist theories that underpin much of population and community ecology (e.g. Lotka-Volterra competition and growth models, and the island biogeography model). It is rare for a text to discuss the deductive versus inductive origins of such models, whether they qualify as theories or even laws, and the extent to which they have universal application. Usually, it remains for the student to decide on such models' utility after they hear a lecture on real-world examples of managing fisheries or elephants. Authors and instructors may justify these models as having "heuristic value" although some would reject such a rationale. For example, Peters[6] has written that, "appeals to heuristic power set a dangerous precedent for scientific judgment since they can become a defence of last resort of bankrupt theory."

The treatment of rationalist models is related to an author's sense of balance between both realism (the existence of general classes) versus nominalism (every case is unique) and knowledge provided by basic research versus that attributed to applied research. In some texts, there are excellent examples of general theory followed by detailed applications of that theory to real systems, but what is lacking is an explicit account of the philosophical leap one makes between the idealized general theory and its empirical testing in a tangible situation.

These are but a few variations that occur within the general, hierarchically structured framework of today's ecology textbooks. We make no attempt, much less claim, to present a detailed analysis or judgment. We only wish to illustrate that fundamental philosophical decisions are made but go unexplained in the construction of these texts. We contend that both the students and the instructor would better understand their subject if these decisions were explicit. This brings us to consider the instructor's attitudes, experiences, and constraints in teaching the introductory ecology course.

General ecology instructors: some conversations

Our second line of inquiry was telephone interviews with a sample of instructors having extensive experience in teaching general ecology across a range of institutions. We acknowledge that this was a mere toe-dipping exercise and a more rigorous, quantitatively valid study might be a useful project. These interviews were structured around a set of questions sent prior to the interview (Box 11.2).

The instructors understood and appreciated the questions we posed, and they espoused enthusiasm for pursuing them in their teaching under the "proper circumstances." By and large, these interviews confirmed our own often frustrating experiences. Inserting philosophical discussion into

---

Box 11.2
Interview questions for ecology instructors

The following questions were sent to colleagues actively engaged in the teaching of general ecology courses:

1) What is the motivation of students to take your course?
2) What are the students' levels of cognitive development with respect to truth?
3) Are students' understandings of truth changed by the course?
4) What is your (the instructor's) cognition of the philosophical meaning of levels of organization?
5) What is the ability of students to accept multiple levels of organization (perspectives) or pluralism more broadly as way of understanding ecology?
6) How do you explain the kinds of ways of learning about truth in nature (epistemology)?
7) How are classical laws of ecology treated?
8) What was the basis of textbook choice?

In most cases optional answers or combinations thereof were offered to the interviewees in the survey document.

---

the classroom was perceived to be counterproductive or worse, given current constraints. Thinking deeply meant covering less material and befuddling students who generally sought clear answers and unambiguous information. But this concern was largely inferential, for if students are confused by multiple perspectives, skeptical about the scientific questions we ask, or dubious about the means by which we learn about the world, it is rarely expressed to the instructors without direct encouragement. Such prompting is uncommon, in that general ecology courses are typically large and do not lend themselves to much open discussion. There is a general body of material characteristic of such courses, largely defined by the textbook, which must be packed into the semester. Student learning is by exam and recorded by grade. As such, it is very easy for these courses to devolve into packaging knowledge as concept-example-problem units to be learned for – and forgotten at the conclusion of – tests. The students know how to play the game as well as we do. After all, behavioral adaptation is a large part of succeeding in college. In our fast-paced, learning environment, conceptual complexity is a nuisance to be minimized, and discussions of values, reality, and truth are luxuries for which there is little time and less interest.

These limitations are not so much criticisms of instructors but properties of American undergraduate education in larger institutions. Understandably, the interviewees had not thought much about the kinds of questions we asked, but they were sensitive to their importance and eager to talk about them. Nevertheless, they saw the reality of their teaching task pretty much as described above, and they were certain that taking time for philosophical issues (e.g. origins and significances of ideas, the subjectivity of concepts, and limits to ways we gain knowledge) would be met with annoyed impatience by students expecting clarity and simplicity, and punished with critical teaching evaluations and diminishing enrollments.

We believe that if instructors attempted to "cover" less material[7] and taught smaller classes more conducive to discussion, that they could and would teach with a more philosophical perspective. From this, we might conclude that the system, rather than the faculty, is responsible for our philosophically deficient methods of teaching ecology. But, of course, we are individually contained in the circular causal loops that make baccalaureate education what it is in America – a parallel to our complicity in defining success in science. We contribute to, even as we are captives of, a system of efficiencies: covering ever more material in increasingly voluminous texts (with CD-ROMs to supplement pages which will never be read in their entirety) in the context of large classes (rather than multiple small sections) that are justified by the economies of scale and time-budgeting. These economies allow us to satisfy other aspects of our job descriptions – particularly research with its hand-maidens of writing (and re-writing) grant proposals, fulfilling promises to ongoing funders, and reporting via publications on results funded by past grants.

Rather than submitting to a cynical view that our culture of science education prohibits change in how we teach ecology, we turn to a deeper examination of the nature of the students encountering our system.

Baccalaureate learning: the nature of our students

To effectively frame the challenge of ecological education, we must take into consideration the motivations of our baccalaureate students. By knowing their motivations we can gain some insight into their initial cognitive and heuristic positions about nature. In our experience, undergraduates are most often drawn to this field through their first-hand experience in the outdoors. In Wyoming, this experience is often hunting and fishing, but there are plenty of students whose interest is rooted in camping, hiking, and – for better or worse – riding ATVs and snow-mobiles (Box 11.3). Researchers differ as to whether a fascination with the natural world is biologically hardwired or socially fostered, or possibly both.[8] What is clear, however, is that the experiences that attract students to ecology persist with regard to their learning style. Few are initially stimulated by theories, computers, or textbooks – they are pragmatists of the sort that John Dewey would recognize. Students generally find that knowing entails doing.

Student survey

To test and augment our understanding of the philosophical positions of undergraduates, we performed a brief survey of a class of juniors and seniors in the early weeks of a field ecology course for which general ecology was a prerequisite (Box 11.4). Almost all of the 25 students responding had career plans in management of renewable natural resources. While the results of this survey are anecdotal, they are nevertheless revealing and largely confirmatory of past impressions.

Our synthetic view of the survey results (Box 11.5) is that undergraduate students seek personal satisfaction and social affirmation by doing

Box 11.3
One avenue into ecology

One of our undergraduate students, Matt Wilson, voluntarily provided this photo along with this statement after a conversation about his fervor for nature and desire to practice an aspect of applied ecology. While some might find the image disturbing, this is the world in which we teach and is the perspective of our students. We suspect that this context is not at all unique to our setting, and it is incumbent on us as professors to see nature through the eyes of our students, as difficult as that image might be.

It's true that hunting has had a critical role in my passion for wildlife ecology. I especially like the connectedness and interactions while hunting not only with the game but the entire ecosystem. As a hunter, I think it is important to actually hunt ... I usually park the truck for the weekend and head out with everything I need on my back. This way I become part of the ecosystem and feel less like an invader.

Box 11.4
A philosophy survey of ecology students

To improve our teaching and gain insight as to how and what students think about philosophy, we gave the following survey to undergraduate students in a field ecology course (N = 25) and to doctoral-level graduate students in the University of Wyoming's Program in Ecology (N = 15).

1. How would you define professional success as an ecologist?
2. Do you think that a broad proficiency in the *philosophy of science* will matter to your success as a professional in ecology?
3. What is the purpose of philosophy?
4. What are the major branches of philosophy?
5. Science philosophy lies within the broader realm of philosophy. What are the central issues in the philosophy of science?
6. Do you think laws or theories that can be used to make predictions about the natural world apply universally, contextually, or in some other manner?
7. Ecologists study both entities (e.g. individual organisms and biotic communities) and processes (e.g. energy flow and co-evolution). List five *entities* that ecologists study in a column below.
8. After each of the five entities you listed, rate the evidence for their "objective reality" (being measurable, tangible, and not dependent on human subjectivity or experience) from 1 (weak evidence) to 5 (strong evidence).
9. Is there more than one way to make a contribution to ecology? For each of the career paths described below, assign your measure of importance of that path to *advancing the science of ecology* and to *advancing the well-being of society* (least importance = 1, most importance = 5).

   As a basic research ecologist, deductively formulate general theory from first principles, axioms, miscellaneous evidence.

   As a basic research ecologist, test general theory with observations and/or experiments of individual systems.

   As an applied research ecologist, use established theory to help resolve local issues of resource and pest management, etc.

   As a research ecologist, collect data and describe exotic, unique, or little-studied situations or phenomena.

Box 11.4 (cont.)

Teach at a 4-year undergraduate college; do small-scale
   research in the summer with undergraduates. Stay current with
   semi-popular and technical articles.

Teach at the junior college level; stay current by reading
   semi-popular articles and limited technical literature.

Teach general science or biological science at the K-12 level.
   Stay current with semi-popular articles.

As an ecologist for a governmental agency, conduct studies on
   natural resources that contribute to management and policy.
   Stay current with technical articles.

As a freelance consultant or member of an environmental
   consulting firm, learn from colleagues and accrued project
   experience as well as reading technical articles in specialized
   journals.

Serve as an in-house environmental officer for a major
   corporation like Weyerhaeuser, Boeing, or Exxon-Mobil.

As an officer or lobbyist with an environmental lobbying
   organization, stay current through attendance at policy-
   directed meetings, and by reading semi-popular articles and
   some technical sources.

10. Rate the following sciences in terms of the complexity of *that
   which they study* (diversity of entities, hierarchical organization,
   spatial and temporal scales, and kinds of interactions) (extremely
   complex = 5, extremely simple = 1): physics, chemistry, genetics,
   ecology, anthropology, sociology.

---

good things for people and nature. However, the students have only a
vague understanding of what might constitute goodness in an ethical
framework. They view philosophy as pertaining to the ineffable qualities
of ultimate meaning and purpose, along with refining the process of
critical thinking. The view that science provides substantive answers
about the world while philosophy fills the gaps with imaginative speculation, including theological musings, was typical. Their understanding
of scientific laws reveals a strong inclination toward pragmatism, albeit
with the well-enculturated parroting of science's universality in the face
of their experience with local contingencies and particular exceptions.
The students' idealism and pragmatism converge in their view of the

---

Box 11.5
Undergraduate students' views of philosophy

Important, qualitative patterns were evident from the survey
responses (see Box 11.4 for complete survey questions):

What is professional success as an ecologist: The students expressed
a sense of idealism moderated by concern for their own forms
of meaning (e.g. "Making a change in the environment for the good"
and "enjoying the work"). The dominant theme entailed doing good
for the natural world and human welfare from which one derives
personal satisfaction and social affirmation.

Does philosophical proficiency matter to your success: The students
generally gave the academic party line: knowledge is good (even when
they don't know what it might be good for). There were some
intriguing insights, such as: "you need to have an understanding of
what you are doing and the reason for doing it."

What is the purpose of philosophy: Undergraduates seemed to hold that
science pertains to the "whats" of the world, while philosophy
addresses the "whys." There was a sense that philosophy served to
refine critical thinking about the world and our claims, as well as
to provide a sense of depth and meaning to life (e.g. "To give reason
and meaning to what we do and why" and "To question everything
and 'dig deep' into 'why?'").

What are the major branches of philosophy: Ethics was the most
commonly identified discipline, but the answers revealed that
students had little idea about the nature of philosophy other than an
impression that it somehow involves both rationality and meaning.

What are the issues in the philosophy of science: Philosophy was
often viewed as a kind of "god of the gaps," providing answers to
questions that are currently inaccessible to science, particularly
matters of cosmological and biological origins (e.g. "Where
everything came from. Where everything is going. How everything is
getting there"). The line between philosophy and theology,
as suggested by earlier answers, was blurred.

Are laws/theories universal or contextual: The students held that laws
are the sorts of claims that really ought to be universal, but they were
deeply skeptical. This tension generated oddly hedged responses such

---

Box 11.5 (cont.)

as: "generally apply universally" and "many predictions will apply universally." They didn't view laws as vacuous, but as useful heuristics that ought not to be relied upon in particular cases.

List five ecological entities that ecologists study and rate the evidence for their "objective reality": The students often listed processes, which suggests that a materialistic realism is so strongly ingrained that nonmaterial existence is difficult to conceive. The most commonly listed ecological entities of highest realism were biosphere, communities, and populations. They were less committed to the reality of ecosystems and habitats.

What is the importance of selected ecological career paths to advancing the science of ecology and the well-being of society: Students indicated that the greatest contribution to science is made by empirical and government researchers, particularly those studying exotic situations. Lobbyists and in-house environmental officers were seen as making the least contribution. As for contributions to society, government scientists were rated most highly while theoretical ecologists were at the bottom. In terms of overall contribution, government and applied ecologists were thought to make the greatest contributions; these are the sorts of positions to which many of our students aspire.

Rate the following sciences in terms of the complexity of that which they study: There was a general, but weak, tendency of increasing complexity from the physical to the biological world. The entities and processes studied by geneticists were judged as being more complex than those of ecology, which suggests that students do not see the former as an element of the latter. Oddly, the subjects studied by the social sciences (i.e. humans and their relationships) were the least complex, but we suspect that students were expressing the belief that the social sciences themselves were not academically difficult.

---

professional path leading to the greatest contributions to science and society – a government ecologist who stays abreast of science and whose research on natural resources contributes to the development of management actions and policy decisions.

We pose these general impressions of the status of undergraduate thought about philosophy and attraction to ecology as a backdrop for what follows as they participate in the American model of education in this area.

Cognitive development and diversity

Perhaps the most pressing issue with respect to incorporating philosophy into undergraduate ecology courses is the cognitive development or readiness of students for abstract thought. The instructors we interviewed consistently expressed doubt that undergraduates would have the ability to wrestle with the complexity embodied in a pluralistic view of ecology operationalized through acceptance of multiple perspectives. We suspect that the worries of faculty reflected not so much the unilateral incapacity of students to engage philosophical concepts but the diversity of abilities and views that students bring to class. With no common starting point, framing a philosophy of ecology is indeed a difficult task. Difficult perhaps, but not impossible if we consider what developmental cognitive psychologists have learned about college students.

The classic work in understanding the mental maturation of students is William Perry's *Forms of Intellectual and Ethical Development in the College Years*.[9] In the last 40 years, various researchers have refined Perry's model such that we now have a reasonably complete account of cognitive development.[10] Although the model is much more finely sub-divided and nuanced than our summary will suggest, the core concept is that students develop through three phases. They typically arrive at a university with an absolutist view of truth and reality, believing that authorities possess knowledge which is transferred to students. It may be the case that science majors come to view authority as not resting in individual experts but existing within "the scientific method" (as if there were a singular procedure) or emerging from the scientific community.

With experience and development, students next enter a relativistic phase. That a liberal arts curriculum – or whatever approximation of this a particular institution has adopted – confronts a student with a stagger-ing diversity of views surely plays a role in this move towards anything being true, right, or beautiful (or nothing being objectively the case). However, the evidence suggests that this developmental transition is not merely an artifact of the American higher educational system, as individ-uals in other cultural settings also tend to exhibit this change.

The third phase is one of a circumscribed relativism or an operational pluralism in which the student acknowledges the existence of multiple views but chooses to adopt and act upon, contingently and fallibilistically, one of these perspectives. The student takes responsibility, and gives reasons for, his or her position. In this phase, an individual might accept that there is progress with respect to knowledge or ethics without this necessitating universal claims or culminating in certainty.

This simplified version of cognitive development has a number of features that are still being explored. For example, whether or not a student who has become a relativist can move back to absolutism rather than forward to pluralism is not entirely clear. While such matters as reversion to an earlier phase are debatable, there are two important refinements of Perry's model that appear to be well established. First, it is clear that individuals do not develop in lock step. That is, in a sophomore-level ecology class, there may be students who will maintain that truth is decreed by authorities, while others hold that truths are purely subjective beliefs, and yet others who understand that various truths may obtain in different contexts. Second, a student's approach to values, knowledge, and reality may well be domain dependent. That is, an individual may be an absolutist with respect to religion and a relativist in terms of material claims, or a relativist with regard to ethics and a pluralist as to aesthetics.

Such cognitive messiness with regard to both inter- and intra-individual development would seem to present an insurmountable challenge for effectively incorporating philosophy in an ecology course. Our survey of ecology students suggested that their philosophical sophistication is highly diverse (Box 11.5), such that an approach that treats them as a monolithic group would be doomed. If one presumes the students are absolutists and therefore presents science as objective truth discovered by experts (*à la* Stephen Weinberg, Box 3.10), then the relativists will object. Conversely, if one presents science as just another way of knowing about the world (*à la* Paul Feyerabend, Chapter 8), then the absolutists will be dismayed.

So, in terms of baccalaureate teaching and learning, where do we stand? Our synthesis of the contemporary state of affairs in undergraduate ecology may be a bit harsh, but it seems that philosophy is expressed – to the extent that it is addressed at all – as a largely unintentional, worrisomely fragmented, potentially incoherent set of implied answers to fundamental conceptual questions. A student of ecology is presented with an ontological/metaphysical foundation of naïve, materialistic realism, an epistemology that hinges on reliable testimony, deduction from first principles and a dose of induction from empiricism, and an axiology that leans heavily on utilitarian ethics. We would propose that making the philosophical framework of ecology both more sophisticated and explicit would be highly beneficial and that constrained perspectivism is a viable solution to the complex problem of embedding philosophy in ecology. This philosophical approach has features that are consistent with both the cognitive capacities of undergraduates and a metaphorical structure that could engage and challenge students. It might even be the case that constrained perspectivism is already lurking at the edge of students' intuitions, as we shall see.

Baccalaureate education: how might philosophy become integrated?

Constrained perspectivism as a viable solution

Many students appear to intuit the problems with absolutism and relativism (e.g. see their responses to the universality of scientific laws (Box 11.5)) and therefore seek a palatable alternative that accords with a more sophisticated approach to a complicated and complex world. We'd suggest that framing ecology via constrained perspectivism is a solution to the false dilemma – and pedagogical obstacle – of absolutism and relativism. The pluralism of constrained perspectivism allows that there is more than one way to be right, but one can still be wrong. And pragmatism, rather than being a colloquial pejorative (i.e. a person "resorts to pragmatism" or one justifies any sort of brutal expediency in the name of pragmatism) can be presented as an honorable philosophical path with rigor, depth, and possibility. Instead of the ordinary notion of pragmatism being "whatever works" students would come to understand that an idea authentically "works" only if it is empirically verifiable as well as socioeconomically and environmentally sound. This means that ecology becomes a vital aspect of pragmatism because the genuine satisfaction of human interests is not possible if the course of action is inimical to environmental sustainability.

Such an approach to the philosophy of ecology might mean that, in the minds of students, science gives up its claims to ultimate power, final truth, and intellectual hegemony. However, we would contend that in this context science becomes more compelling, not less. Ecology gains a profound relevancy to the lives of students – a most important consideration (Box 11.5) – when it is woven into the human experience. Ecologists are seen as engaging in the process of world-making through experientialism (see Chapter 7). This view is more engaging than subjectivism, in which humans become All Powerful with respect to defining beauty, truth, and reality, but there is no better or worse version of the world so the exercise becomes degenerately narcissistic. Experientialism is more promising than objectivism, in which humans are Impotent Mirrors of the world who slavishly and meaninglessly attempt to reflect reality by removing ourselves from experience.[11]

By casting the philosophy of ecology in terms of constrained perspectivism, students within any of the developmental phases are both reassured and challenged. The absolutist can find security in there being a real, constraining world. But such an individual is challenged by the importance of contexts and interests in shaping how we engage and understand this world. The relativist can find safety in the contingent domains and human satisfactions that parse reality in subjective ways. But this student is challenged by the existence of an objective world that pushes back via empirical verification or refutation of our ideas. For the pluralist, constrained perspectivism offers a formalization and refinement of what might otherwise be a vague and muddled intuition about a workable framework somewhere between the two extremes. Pragmatism, and the particular formulation that we've developed as constrained perspectivism, offers students a rich, rigorous, and nuanced conceptual framework with immediate application not only to understanding ecology in an intellectual sense, but to doing ecology in an operational sense. This practical feature of philosophy will likely engage students with an affinity for hands-on, applied learning.

Perhaps the greatest concern about constrained perspectivism that instructors of general ecology courses might raise is the seriousness with which subjectivity is treated – human interests have a necessary and legitimate, if not defining, role in science. In response, we emphasize that perspectives are constrained. There is a world that informs us of the viability of our ideas when they are put into action. When our ideas give rise to actions that produce suffering, filth, and ugliness, we have a failed concept. While such stark terms may not always be the case in ecology, there can be no doubt that we have made mistakes. For example, consider our earlier case of discovering whether nitrate would be lost based on the experiments we performed in forests across the USA (Chapter 6). Although unspoken among the participants, some of us were eager to show that "our" systems would (or would not) follow the Hubbard Brook experience based on previous convictions we had as to how particular systems worked. Another unspoken bias for some was the wish and expectation that "ours" would show the biggest and fastest response to the experimental treatment. Alas, the systems did what they did and we quietly had to revise our convictions about the functional character of these systems.

A further worry might pertain to the social nature of science. Does it not undermine the credibility of ecology to tell students that the determination of truth is a communal process? We'd contend that exactly the opposite is the case. Exposure of ideas for the rational review and empirical testing by others sets science apart from other ways of knowing. Constrained perspectivism reinforces this notion in reminding students (and scientists) that if we do not listen to others or if we categorically exclude evidence produced by alternative methods, we close off essential dialogue. Intersubjective objectivity within a domain means that an intellectual community checks the slide into nihilism. Because we are (or can be) rational and share experiences (in relevant, although nonidentical, ways), ecologists are able to compose an internally coherent and profoundly useful system for understanding and engaging the world. Indeed, it seems that humans, including scientists, are most dangerous when they act in isolation, without social constraints on their world-making.

Techniques for introducing philosophy
to undergraduates

The matter of how best to present philosophy in courses or textbooks of ecology is challenging. Based on our interactions with undergraduates (Box 11.1), three issues must be considered. First, students are attuned to relevancy and often are unwilling to engage a topic for a long period of time or with substantial intensity if there is no evident application to their lives (Box 11.5). This is why it is often difficult to interest students in the history of science. Unless there are fairly direct ramifications for their immediate concerns, the story of ecology's development, conflicts, and characters is deemed irrelevant. Such could well be the fate of philosophy if we fail to attend to how students process information. Our experience suggests that a vertical approach may be a weak strategy (e.g. "For the next week we'll learn about philosophy" or "Chapter 3: The Philosophy of Ecology"). Rather, an integration of philosophical concepts throughout a course or text is likely to be much more effective (e.g. "Given what we've learned about ecosystems, do you think that they actually exist in the world?" or "Let's consider what counts as evidence in justifying our claim that we know the climate is changing.").

Next, students are often resistant to didactic approaches, particularly when they have existing views on a subject. This can be a problem for teaching ecological concepts, given that many students have preconceived views about the way in which natural systems function. Bar room biology and hunting camp ecology pronounce that without predator control, wolves or mountain lions will diminish elk populations to extinction. The challenge is even greater when raising questions about beauty, truth, and right, for all people have beliefs (often fiercely held) about such matters. As such, a communal approach based on tactical questioning seems to offer substantially more promise than a pedantic lecture as a means of both engaging students and introducing ideas that may challenge their existing beliefs. Socratic questioning is a particularly promising, but not easily mastered, entrée into such matters.[12] At some point it is incumbent on the instructor to impose synthesis, direction, and cohesion that provides a framework of constrained perspectivism as a philosophical model for ecology. However, it seems likely that in a typically diverse class of undergraduates, there will be students who advocate and refute both objectivism/absolutism and subjectivism/relativism thereby providing the instructor with the raw material for presenting constrained perspectivism as a viable alternative.

Third, the matter of substance and style comes into play. Just how much philosophical content is optimal and how much open-ended discussion is valuable with respect to authentic learning? A practical approach may be to work within the framework described by Diane Ebert-May and Janet Hodder.[13] Based on considerable experience, the education program they describe carefully sets an educational framework for the instructor and establishes intellectual standards and guidelines from the first day of class so that students immediately know this experience is going to be different. The method combines reading relatively synthetic papers having applied consequences – an appeal to student interests – followed by a structured form of critical analysis. The system promotes consideration of multiple epistemological methods for different kinds of problems – an introduction to pluralism at this level. It seems to us that constrained perspectivism could underlie the entire approach, although it appears to be revealed through practice rather than by explicit argument. It is not evident to us that elements of ontology, metaphysics, and axiology or the importance of interest satisfaction are embedded in the method, but it is obvious that some discussion of these topics easily would fit into this approach.

The way in which a new concept is framed is critical to how the idea is received and integrated into an individual's existing worldview. As such, the term "constrained perspectivism" is hardly the best way of introducing this philosophical system. This phrase, however descriptive it may be once one has explored the attendant ideas, is not likely to initially make sense to students. What is needed is a metaphor that allows the new concept to be connected to a familiar context or experience.[14] There may be other and better metaphors, but we'd suggest that the concept of a lens has significant potential. Virtually all students will have peered through a pair of eyeglasses or played with a kaleidoscope; those in the sciences will have experience with lenses in microscopes and telescopes and many will be familiar with aspects of remote sensing (via Google Earth, if in no other way). Framing constrained perspectivism as a set of lenses entails several useful and important qualities of this philosophy, including the existence of an observer (subject) and the observed (object), the inevitability of distortion, the active choice of a lens based on interests, the absurdity of a view being right or wrong in an absolute sense, the effect of scale, etc. (Box 11.6).

Box 11.6
The lens of constrained perspectivism

Understanding constrained perspectivism via the metaphor of lenses is a potentially engaging and effective means of capturing the unfamiliarity of philosophy in a familiar framework. The concept of an instrument with multiple lenses (a compound microscope, for example) entails a number of features that are shared with the philosophy of constrained perspectivism (CP):

Lenses are useful tools for scientists → CP is a conceptual instrument that is handy for ecologists.

Lenses are used by observers to examine the real world → CP admits of there being an objective reality.

Lenses provide a view that is neither purely created (we don't imagine what we see) nor entirely discovered (there are no perfectly clear, undistorting lenses) → CP argues that we are world-makers crafting what we perceive within objective constraints.

Box 11.6 (cont.)

Lenses take many forms and the one we choose depends on our interests → CP maintains that we select a perspective as a function of our desires.

Lenses don't provide a right or wrong view of the world but a useful one → CP links our choice of scale and complexity to whether or not this perspective meets our needs.

Lenses can filter wavelengths to highlight certain aspects of an entity and obscure others → CP proposes that we possess partial truths but never the whole of reality.

Lenses can deceive the observer if one forgets that a particular filter or magnification is at work → CP warns us that we must be keenly aware of our choice of perspective or risk mistaking a particular view for objective truth.

Graduate student learning: pursuing a career in ecology

Why do they commit?

Presuming that a student works his or her way through the undergraduate coursework in ecology, the next step is crucial. Deciding to pursue this field as a career is a major commitment, and to continue our analysis of teaching we must attempt to understand why students want to become professional ecologists. Again, through our conversations with students contemplating this decision, we can suggest the most and least likely explanations. Within our academic, socioeconomic, and geographic context, students usually give a values-based rationale and express an intention to "make a difference" in science or society. The aspiration to lead a meaningful life in which humans, species, places, or all of nature is better off for one's work strikes us as just the sort of foundation that can sustain a life-long engagement with ecology. With almost equal frequency, it seems that students are motivated toward an ecologically based career through appreciation of the beauty of the environment together with admiration of, and a kind of alliance with, wildlife (even, perhaps especially, if they hunt) that is sometimes expressed as "love" of nature. Conversely, very few students express an interest in discovering universal, objective truths about the natural world. One has to wonder whether this might more often be the case with students in physics, for example. In any case, we have found that motivations for committing to a career in ecology more often stem from aesthetic and ethical influences than from intellectual ones.

To the extent that students refer to ecological principles or theories, they seem to be more often understood as potentially useful tools of problem-solving rather than intrinsically valuable jewels of knowledge. But for many students, the basic-applied distinction seems to be lost. As budding pragmatists, it appears that what matters to them is how an idea can be converted into some practice or policy.

Students often reflect on both their past experiences and future aspirations. Childhood stories frequently emerge as explanations, and their desire to have such meaningful experiences in the future is evident. This Peter Pan quality of ecology in which one never has to entirely grow up – or at least those encounters with nature that provided meaning in the past might still be a source of inspiration in the adult world of science – seems to be an important element in ecological education. Some cynically and ruefully admit they want to turn their hobby or recreation into a job. However, we suspect there are deeper emotional underpinnings to this attitude, one of which might be an escape from the more demanding, competitive and rapidly changing global scene that requires greater intellectual and attitudinal adaptation than they wish to face. Just as worldly, urbane sophisticates find solace in timeless, classical music, these students find refuge in the natural world away from the maddening crowd.

Graduate education in ecology seems to be a hybrid of the master-apprentice model for fostering craftsmanship and the academic model for cultivating a life of the mind. We first consider the master-apprentice model, an approach that is highly compatible with the philosophy of constrained perspectivism and its emphasis on the pragmatic qualities of ecology-as-practice. If truth is the quality we ascribe to ideas that are verified through the consequences of their implementation, then it seems reasonable to perceive graduate education as providing hands-on practice for solving problems within a particular context. In this context, Richard Karban and Mikaela Huntzinger's *How to Do Ecology*[15] is understandably akin to a mechanics manual. But as much as these authors try to stick to the nuts-and-bolts of vocational ecology (and winning the game of science), they can't resist the question: "Why do ecology?" In their recommendations for making strategic decisions about immediate and mid-term goals, they insert the following parenthetical aside concerning the importance of formulating long-term goals: "If you don't believe this, talk to some burnt-out researchers late in their careers. Some people never bothered to stop and figure out what they really valued and wanted to accomplish for themselves." Although the advice seems to endorse a kind of narcissistic axiology without duties to others, at least there is recognition that proximate benefits are insufficient for a meaningful life in science. We'd suggest that social obligations constitute a rather more viable basis for sustaining a long-term commitment, and there appears to be a growing recognition that "publicly funded science has a duty to dedicate resources toward overcoming [global environmental] challenges."[16] Indeed, Donald Strong has argued that, "Whereas ecology is science and environmentalism sometimes is and sometimes isn't, the latter is necessary for the former ... Why should we not advocate for protection of the environment in our professional capacity?"[17]

Turning now to the academic model of graduate education, Karban and Huntzinger's admonition might serve as a reason for delving into other philosophical questions. These epistemological, ontological, and metaphysical issues have the potential to eventually undermine an ecologist's world, particularly if that world is built on a hard-driving realism, absolute truths, certain knowledge, and universal laws. As such, the academic model is also supportive of explicit philosophical engagement with the substantive, conceptual challenges in ecology. It may well be as disturbing to wake up one day wondering if ecological communities are fictional as it would be to doubt an arcane sliver of science was worth an investment of one's life. In principled defense of including philosophical discourse in graduate education, we'd suggest that if a Doctor of Philosophy is anything other than an historical vestige, then it is reasonable to expect a person possessing a Ph.D. in ecology to understand the philosophical foundations of his or her science.

The status of philosophy in the graduate experience

In order to probe the extent of knowledge and concern for philosophy on the part of graduate students, we applied the same survey described in Box 11.4 to graduate students early in their careers in a Ph.D. program in ecology at the University of Wyoming. Numbers were small but, based on other experiences, the answers were indicative of graduate student knowledge at our institution.

The synopsis of survey results (Box 11.7) suggests that graduate students, even those at the doctoral level, appear to have little better understanding of philosophy than do undergraduates in a general ecology course. Given that very few graduate students have had an introductory

Box 11.7
Ph.D. graduate students' views of philosophy

Important, qualitative patterns were evident from the survey responses (see Box 11.4 for complete survey questions):

What is professional success as an ecologist: Responses reflected a cog-in-the-machine understanding of the role of an individual scientist as exemplified by phrases such as "[providing a] piece of the bigger picture." Although seeing their research as adding a brick to the edifice of science, there was often a mitigated idealism or a latent aspiration toward deeper purpose (e.g. "a meaningful contribution to conservation").

Does philosophical proficiency matter to your success: The students were either skeptical (e.g. "I'm dubious" and "It would increase understanding, but not success") or expressed the diffuse belief that liberal education was an undefinable, mysterious virtue, as exemplified by this response: "Most likely, although I don't know how."

What is the purpose of philosophy: Most students expressed a vague sense that philosophy was a method for reflective, even critical, thinking (e.g. "To question and define theories, concepts, and ways of doing things"), but there was little understanding that it constituted a body of knowledge in its own right.

What are the major branches of philosophy: Epistemology was most often mentioned, with logic, ethics, and metaphysics occasionally identified. Many students listed topics such as science, art, medicine, law, etc. Perhaps the most telling response was, "Something I intentionally avoided as a student thus far."

What are the issues in the philosophy of science: The responses most frequently concerned epistemology (e.g. "the validity of the scientific method as a way of obtaining knowledge"). There were occasional allusions to ethics such as vague references to "Man ... messing with what is beyond him," but no further axiological or ontological/ metaphysical concerns.

Are laws/theories universal or contextual: Most students were dubious of universals but seemed cognizant that laws were supposed to make such claims resulting in some odd expressions: "Universally, but

---

Box 11.7 (cont.)

probably with exceptions." There seemed to be a general sense that ecological theories are context-sensitive or domain-specific.

List five ecological entities that ecologists study and rate the evidence for their "objective reality": The five most commonly listed entities spanned the scale from the individual to the ecosystem. Organisms were taken to be objectively real (avg. 5.0), while the students were generally anti-realists about ecological entities, with populations, communities, and ecosystems receiving average scores from 3.5 to 3.8. Species were taken to have the least objective reality (2.0).

What is the importance of selected ecological career paths to advancing the science of ecology and the well-being of society: The students saw the greatest contribution to science through basic research and the greatest importance to society through teaching. There was an inverse relationship between careers that benefited science and those that contributed to society. However, if one wanted to advance both science and society through a single career, applied research was the optimal path.

Rate the following sciences in terms of the complexity of that which they study: The students viewed ecology as having the greatest complexity. Genetics and the physical sciences (physics and chemistry) were perceived as least complex, while the social sciences (anthropology and sociology) were intermediate. It would seem odd that sociology or anthropology, which arguably includes much of ecology along with human behavior, would be less complex than ecology. However, we suspect that students thought that the social sciences themselves were less difficult to understand than ecology in terms of concepts, models, and mathematics.

---

class in philosophy, let alone advanced coursework in the philosophy of science, this ought not to be surprising. Indeed, if doctoral candidates learned elements of philosophy as part of a liberal arts education, they are even more distant from these experiences than advanced undergraduates or masters students (Box 11.5). The most knowledgeable responses obviously came from students who had taken our introductory class in ecological history and philosophy – coursework in these realms demonstrably improved understanding.

Our general experience with teaching a required graduate-level course on the philosophy and history of ecology in the University of Wyoming's Ph.D. program is that most students are somewhat resistant and most on-looking advisors are rather dubious. The central concern seems to be that coursework is an impediment to completing a dissertation, and classes that do not directly impinge on the student's research are a questionable use of valuable time (having advised dozens of graduate students, we are sympathetic to this view). Appeals to the ideal of authentic scholarship, the integrity of the field, the holistic development of a scientist, and putting the Ph. back into the Ph.D. don't carry much weight when a grant-funded project depends on the graduate student completing his or her research. So, one might ask, can the standards of pragmatism be used to defend the inclusion of philosophy in the education of graduate students? Are genuine human interests satisfied if Ph.D.s in ecology are philosophically literate? (Box 11.8).

We contend that understanding the philosophy of ecology is a valuable use of a graduate student's time. In defense of this position we offer a set of arguments that may be more or less compelling depending on what the reader takes to be the purpose of graduate education, but we hope that at least one of these lines of thinking will be persuasive.

Box 11.8
Educational models

If we stipulate that philosophy should be part of ecological education, there are at least three models that warrant consideration.

Liberal education model: Undergraduates take an introductory course in philosophy and perhaps a further course in philosophy of science. Graduate students learn the philosophy of ecology from their mentors (who learned it from theirs).

Cons: Most institutions lack the faculty in philosophy to teach all science majors a general course, let alone one in the philosophy of science. Few programs in ecology have faculty prepared to teach the philosophy of their discipline.

Pros: Such a "building block" approach would sustain a continuously refined understanding of philosophy, providing both introductory breadth and subsequent depth.

Box 11.8 (cont.)

Remediation model: Because graduate students do not, in fact, have previous coursework in philosophy, either ecology or philosophy faculty provide both the essential foundation and the more disciplinary-specific aspects of this field.

> *Cons*: Most texts and courses in the philosophy of science are devoted to an idealized version of physics, so the responsibility for addressing philosophy would fall to ecology faculty. Most ecologists require their own remediation to teach philosophy.

> *Pros*: Having faculty and graduate students co-learning philosophy can represent an ideal educational setting and demonstrates to students the importance of intellectual curiosity and humility.

Mixed model: All incoming graduate students are expected to have had a course in philosophy (if not, remedial coursework is required), and ecology faculty teach the philosophy of their discipline.

> *Cons*: Ecology faculty would need to become familiar with general philosophy and well versed in the philosophy of their field, which would require an admission that we often don't know what we don't know.

> *Pros*: The students would be assured of a philosophical exploration via a guide or co-learner who can assume foundational knowledge and shares their disciplinary interests and concerns.

Arguments for requiring an operational philosophical knowledge of graduate students

We have seen students, at various points in their graduate education, experience erosion or collapse of their worldviews. This cognitive dissonance can arise through understanding that a particular entity or process does not objectively exist, that an arcane line of research is not contributing to a world in need of solutions, that solving puzzles and arguing about abstractions is not fulfilling, that science is ultimately "impure" and profoundly political, or that nobody, including oneself, is any happier for what one is doing.[18] Many, if not all, of these realizations arise from a hard-driving realist position that can no longer be sustained in the face of experience. We suggest that inoculating students against these existential crises is a far more effective, not to say humane, approach to education than hoping that an individual will never confront such trying times.

In an early review of the manuscript on which this book is based, a reviewer from what we were told was an "elite Eastern university" castigated us with the contention that: "The basic message [of this book] is needlessly discouraging, not one that motivates graduate students to do better science." The reviewer was probably right, if "doing better science" meant acquiring the means to generate conventional publications. Nothing else in this particular review suggested a concern with quality, originality, or usefulness. As such, the implication of this comment was that mentoring for academic productivity is the entirety of our obligation to students. By this account, our duty is the same as that of a virus – self-replication.

Perhaps it is the case that providing future ecologists with an antidote to the cynicism that is likely, if by no means certain, to arise can be discouraging in the present. But then, hoping that a student will sustain a lifelong belief in the myth of science as progressing steadfastly toward absolute, objective truth (or that a youngster will never realize the problems of a literal Santa Claus) seems profoundly disrespectful of the cognitive abilities of those who we are supposed to be educating. Even if a student (or a mentor) has managed to hang onto an absolutist or authoritarian view of truth within the domain of science, it seems untenable to wager that such a psychological state is indefinitely sustainable. Rather, we have a duty to prepare the individual to constructively engage potentially compelling alternatives.

Another pragmatic argument for including philosophy in graduate education is the legitimate interest that ecologists have in the intellectual integrity of their science. We are surely in need of highly competent and productive technicians, capable of meticulously following protocols and even adapting these rules to meet unexpected conditions. But ecology is more than a matter of generating publishable data and fundable proposals. Without some scientists – and perhaps all of those holding a doctorate – being able to raise deep questions about the values, knowledge, and realities that comprise and limit ecology, there is little hope for truly new and revolutionary theories (Box 11.9).

Another consideration for the value of philosophy to students of ecology is that such understanding has the potential to foster research (and teaching) that is of a substantially higher quality than that generated by the merely skilled practitioners. We can't contend that philosophically sophisticated ecologists will publish more papers and acquire more grants than their more narrowly trained colleagues. Indeed, just the opposite

---

Box 11.9
The parable of the performer

Imagine that there is an opera singer with an excellent voice. She can reliably hit all of the notes on the score, she never misses a cue, and her diction in Italian (she specializes in Puccini) is exquisite. But she knows nothing of music theory, is utterly clueless about the Italian language, and has no understanding of the historical or cultural context of La Bohème, Tosca, or Madama Butterfly. As such, she is a technical virtuoso but is unable to provide an original or creative interpretation of the parts that she sings.

Is she truly a musician? Or is she, instead, a performer of music? In German, there are different words for these two possibilities. A *musiker* has a deep understanding of music, while a *musikant* plays music – perhaps quite well, but without grasping the cultural, historical, aesthetic, and conceptual context.

A reason why philosophy of science is important is that, although no such words exist, there is an important difference between a *scientifiker* and a *scientifikant*.

---

seems possible. But is there a tremendous shortage of data, papers, and projects in ecology? Or are we lacking in original ideas, incisive syntheses, and provocative investigations? Ecology (like most, and perhaps all, sciences) is drowning in information, but hungry for knowledge and starving for wisdom.[19]

The pragmatist is not concerned with satisfying just any human interest. As noted earlier in this book, a pragmatist would not consider providing heroin to an addict as a defensible action even though in doing so a human desire is verifiably and empirically fulfilled. What concerns us as teachers or mentors are the genuine interests of our students. A part of education is the cultivation of the desire for the right sorts of knowledge. Of course, detailing what constitutes such a desire is a tremendous task in light of the enormous variety of mentors, but we take this to mean at least an aspiration to knowledge that provides an abiding source of meaning and motivation for a fulfilling life. "Operational philosophical knowledge" in this sense means that one can function at a high level of understanding, commensurate with an active, fulfilling, and sustainable life of the mind. And insofar as this is the case, we would maintain that an understanding of philosophy substantially contributes to long-term satisfaction consistent with the human potential of many (not all, as we shall see) ecologists. This brings us to our final point.

A lifetime's devotion to ecology has its virtues

Although changing careers is not necessarily a problem (the needs and interests of individuals may well change over time), there is much to be said for having a cadre of senior scientists with extensive and varied experiences. We would contend that a professional life dedicated to ecology is most likely to unfold when a scientist has a deeply considered axiology (the idealism of the undergraduate appears to have been substantially moderated by the time one is a doctoral student; Boxes 11.5 and 11.7). If, as we argue, inquiry is interest-based, then becoming keenly aware of, and consciously cultivating, these interests seems a plausible antidote to burn-out and cynicism.

In our conversations with graduate students, it often emerges that they evince a public and a private set of values to justify their research. In the former context, they frequently offer ethical – usually utilitarian – accounts of how their project will benefit humanity (e.g. "my research will help mitigate the damage of global climate change to ecological processes that sustain human well-being"). The linkages to the public good are often tenuous and convoluted, but students are well-practiced in providing such speculative connections through their reading of scientific papers and preparation of research proposals. Upon further reflection and with the safety provided by a questioner who welcomes honesty, graduate students often evince that a deep sense of beauty is at the core of their personal values concerning ecology. Their reluctance to admit as much is often rooted in the erroneous belief that aesthetics is purely a matter of subjective taste or mere opinion. As such, students are surprised – and extremely pleased – to learn that aesthetics is an intellectually rigorous field of study and an entirely legitimate basis for one's values. That one can justify an abiding devotion to ecology through a sense of beauty that is richly informed by science[20] is a revelation of the sort that may have the potential to foster life-long commitment.

Some ecologists might object on experiential grounds to our contention that philosophy ought to be explicitly included in graduate education. After all, one might argue, there are any number of highly productive ecologists who persist in their research without any evident grasp of the underlying philosophical principles. But this is like arguing that because there are many musical performers who make a fine and consistent living by technical competence without understanding music theory, history, etc. we ought not to teach the cultural and conceptual framework of the arts (Box 11.9).

How might philosophy be incorporated into the graduate experience?

Assuming that the reader finds at least one of these lines of argument sufficient to justify the inclusion of philosophy in the education of graduate students, we are now left with a couple of logistical matters. First, how important is philosophy for masters versus doctoral students? One might argue that the M.S. is a kind of performance degree while the Ph.D. requires conceptual sophistication. If so, and assuming that an incoming masters student has a basic understanding of philosophy from undergraduate coursework, then perhaps further development is not required. However, greater reflection on one's philosophical foundations – particularly axiology – might be valuable whether or not the student intends to pursue a doctorate.

Second, as with undergraduate education the question of a vertical versus horizontal approach is relevant. On the side of verticality, a single course that focuses on the philosophy (and perhaps history) of ecology makes clear that faculty takes very seriously the importance of this intellectual venture. But this approach risks the "Wednesday afternoons are when I think about philosophy" problem in which deep questions are relegated to a particular experience within graduate education. In favor of horizontality, integrating philosophy across courses (and ideally within the research proposal, preliminary exam, thesis/dissertation defense, etc.) makes such thinking a constant practice. However, this approach risks missing the depth of reading, discussion, and learning that is required to gain a substantive grasp of philosophy. As with so many such matters, the sensible answer is probably "both." A concentrated exploration of philosophical concepts early in a graduate student's education, followed by frequent applications of this knowledge in various courses and research settings would be optimal.

A philosophical standard for graduate students

In summary, there would seem to be much to gain and little to lose if graduate education included philosophical inquiry as to what is valuable, known, and real (matters that are now superficially understood by students, Box 11.7). That a plurality of perspectives constrained by the real world is a useful framework for such considerations is, in the end, less our point than the importance of the inquiry itself. Whether a student adopts an absolutist, relativist, or pluralist view of science is not so much the issue as that the view be intentionally developed and thoughtfully defended. That is, as much as we favor constrained perspectivism it would be inconsistent and disingenuous to suppose or demand that students adopt or arrive at this philosophical position.

A rich pluralism of ecology programs such that students can find a system which accords with their desires would seem to be the logical outcome of a pragmatic analysis of graduate education. But there is a catch. Pragmatism attends to the genuine needs of people, not merely their desires for satisfying proximate wants, such as yearning for fatty foods (a desire contrary to the authentic interests of people for long, healthy lives). As such, there may be nonnegotiable standards which necessarily accord with the genuine needs of students. This seems to endorse a kind of paternalism, or at least a commitment to what we understand as the core of a graduate degree in ecology. For if faculty don't "know more" about what constitutes their discipline and we are simply here to provide for our "customers" (the business model of education), then education becomes a kind of fast-food enterprise with administrations pleased to have sold lots of intellectual French fries. We are willing to don the mantle of paternalism – or at least to commit to our understanding of the integrity of a university education. There is at least one standard of philosophy to which we are willing to hold all graduate students in ecology.

The goal that we are willing to advance in an uncompromising fashion (at least within the domain of graduate education) is that the students should be keenly aware of and explicitly responsible for the philosophical views they hold. Appeals to cultural pressures, social norms, economic expediencies, and other external factors are not acceptable foundations for advanced students in the science of ecology. Here we borrow from existentialism, a branch of philosophy most succinctly summarized by Delmore Schwartz: "Existentialism means that no one else can take a bath for you."[21] In this light, we insist upon what the existentialist philosophers call authenticity:[22]

> What [authenticity] means can perhaps be brought out by considering moral evaluations. In keeping my promise I act in accord with duty; and if I keep it *because* it is my duty, I also act morally (according to Kant) because I am acting for the *sake* of duty. But existentially there is still a further evaluation to

be made. My moral act is *inauthentic* if, in keeping my promise for the sake of duty, I do so because that is what "one" does (what "moral people" do). But I can do the same thing *authentically* if, in keeping my promise for the sake of duty, acting this way is something I choose *as my own*, something to which, apart from its social sanction, I commit myself.

Our ontological, metaphysical, and epistemological commitments ought to be made in this same matter, if we are to be responsible and authentic ecologists.

Implications for "post degree" learners

The education of an ecologist does not stop with the granting of a terminal degree. Faculty and other professionals continue to learn about new methods and procedures, novel theories and concepts, and recent findings from allied and divergent fields. Along with adapting one's research and teaching to these unfolding sources of knowledge, the ecologist might also be expected to continue confronting philosophical issues. In our experience – being typical ecologists without formal training in philosophy as graduate students – as we struggled with these deep issues, problems with the nature of value, knowledge, and reality emerged. From conversations with colleagues, it is clear that we are not alone in having run head long into our flimsy understanding of truth. Like many other scientists, we'd placidly assumed that our research contributed to a continuous convergence on absolute reality. The realizations that objective truth is a mythic notion ("mythic" being a conceptual description, not a pejorative attack[23]), that science is infused with subjectivity, and that we can't know anything except through our limited senses and idiosyncratic experiences, were deeply troubling insights.

Grasping the contingent and interactive nature of science opens three paths. First, one could wall-off this nagging doubt, denying that it is relevant or problematical to life as an ecologist. This partitioning of the philosophical from the scientific is psychologically plausible. We know of a fellow scientist who is a creationist on Sundays and studies the evolutionary responses of organisms to environmental toxins on weekdays. If this dividing of one's religious and scientific beliefs into incoherent realms is possible, then surely keeping philosophy from muddling science is doable. Second one could interpret this failure of objective truth as a dirty secret and decide to play the game of science. For the cynical ecologist, this might mean scoring grants, keynote invitations, and other badges of success in a social contest of unapologetic self-promotion. Third, one might – as we did – explore how there can be genuine meaning and substantive progress without certainty. For us, this led us on convergent paths to constrained perspectivism vis-à-vis pragmatism. We came to understand there is a world "out there" about which we can know things through its capacity to push back on our beliefs in response to the actions that we take. But the hope of attaining certainty in some singular and noncontextual manner has been abandoned. Science may not be a Santa Claus that delivers truth to good researchers, but this is not to say that there is nothing that can be done in the world that makes life better or mitigates suffering.

"Philosopause," a term apparently coined by Donald Griffin, has taken on a rather negative connotation in referring to the tendency of over-the-hill scientists to speculate about the deep meaning of their careers through self-aggrandizing autobiographies.[24] But we'd suggest a rather more charitable interpretation may be in order. The definition offered by Anne Soukhanov, suggests that philosopause comes at a time when, "a researcher, weary of or frustrated by rigorous laboratory-based science, begins to look for nonscientific, philosophical explanations."[25] Even this is hardly a laudable notion and oddly implies that scientific and philosophical endeavors are somehow independent, even opposing. Setting aside this implication, Soukhanov's take on the maturation of a scientist points out that philosophical reflection can be induced by a weariness or frustration – perhaps not so much through an exhaustion with "rigorous" science but, we'd suggest, through a realization that the promises of science that were implicitly made to us as students are not explicitly realized in our labors as ecologists. Perhaps this understanding comes with the opportunity for deep reflection, some time after the "productive" flurry associated with tenure and promotion, when one has the luxury to shift from quantitative to qualitative measures of success.

Whatever the events or cognitive developmental pathways that lead to an ecologist's professional angst, it is our sense that the education of scientists should prepare individuals for a crisis of faith. By providing our students with a background in philosophy, by cultivating a sense of deep inquiry, by attending to the values and interests – the fundamentally human interests in self, others, and the world – that ultimately motivate their work, we can prepare them for whatever may come. If there is no "dark night of the soul" during which they are forced to confront their doubts, then the students are no worse off for having a richer understanding of the nature of science. There is much to be said, even by a pragmatist, in favor of the *scientifiker* (Box 11.9). In the end, we hope that constrained perspectivism meets the standard of pragmatism, that this philosophical framework for ecology proves useful to the students of ecology, including those who have finished their formal education.

Endnotes

1. The Newseum website, newseum.org/yesvirginia/, accessed October 13, 2008.
2. One of our professorial interviewees has discovered that some students having the option to choose ecology from an array of intermediate biology courses, elect ecology because of its quantitative nature.
3. Gould, S. J. 2002. *The Structure of Evolutionary Theory*. Cambridge, MA: Harvard University Press, 576.
4. Wilkinson, D. M. 2006. *Fundamental Processes in Ecology: An Earth System Approach*. Oxford, UK: Oxford University Press.
5. Bormann, F. H. and G. E. Likens. 1967. Nutrient cycling. *Science*, **155**: 424–429.
6. Peters, R. H. 1991. *A Critique for Ecology*. Cambridge, UK: Cambridge University Press.
7. Instructors also may be under some compunction to prepare students for the Graduate Record Exam that contains questions over a variety of topics, some of which may be missed in a more focused presentation of ecology. There may also be departmental expectations, course prerequisites, and other "standards" that drive instructors toward breadth and away from depth.
8. Wilson, E. O. 1986. *Biophilia*. Cambridge, MA: Harvard University Press; Louv, R. 2008. *Last Child in the Woods: Saving Our Children From Nature-Deficit Disorder*. New York: Algonquin.
9. Perry, W. G. 1970. *Forms of Intellectual and Ethical Development in the College Years: A Scheme*. New York: Holt, Rinehart and Winston.
10. Chandler, M. J. 1987. The Othello effect: an essay on the emergence and eclipse of skeptical doubt. *Human Development*, **30**: 137–159; Kardash, C. M. and K. L. Howell. 2000. Effects of epistemological beliefs and topic-specific beliefs on undergraduates' cognitive and strategic processing of dual-positional text. *Journal of Educational Psychology*, **92**: 524–535. Kitchner, K. S. and P. M. King. 1981. Reflective judgment: concepts of justification and their relationship to age and education. *Journal of Applied Developmental Psychology*, **2**: 89–116; Kitchner, K. S. and P. M. King. 1989. The reflective judgment model: ten years of research. In *Adult Development III. Models and Methods in the Study of Adolescent through Adult Thought*, eds. M. L. Commons, C. Armon, L. Kohlberg, F. A. Richards and J. D. Sinnott. New York: Praeger, 63–78; Kitchner, K. S. and P. M. King. 2002. The reflective judgment model: twenty years of research. In *Personal Epistemology: The Psychology of Beliefs about Knowledge and Knowing*, eds. B. K. Hofer and P. R. Pintrich. Mahwah, NJ: Lawrence Erlbaum Associates, 37–62; Jacobs, J. E. and P. A. Klaczynski, eds. 2005. *The Development of Decision Making in Children and Adolescents*. Mahwah, NJ: Lawrence Erlbaum Associates.
11. Lakoff, G. and J. Johnson. 1980. *Metaphors We Live By*. Chicago: University of Chicago Press.

12. To develop the ability to engage in Socratic questioning as an approach to learning, the reader might consider: Paul, R. 2006. *The Thinker's Guide to the Art of Socratic Questioning*. Dillon Beach, CA: The Foundation for Critical Thinking; Phillips, C. 2002. *Socrates Cafe: A Fresh Taste of Philosophy*. New York: Norton.
13. Ebert-May, D. and J. Hodder, eds. 2008. *Pathways to Scientific Teaching*. Sunderland, MA: Sinauer Associates, Inc.
14. Lakoff and Johnson, *Metaphors We Live By*.
15. Karban, R. and M. Huntzinger. 2006. *How to Do Ecology: A Concise Handbook*. Princeton, NJ: Princeton University Press.
16. Cabrera, D., J. T. Mandel, J. P. Andras, and M. L. Nydam. 2008. What is the crisis? Defining and prioritizing the world's most pressing problems. *Frontiers in Ecology*, **6**: 469–475.
17. Strong, D. 2008. Ecologists and environmentalism. *Frontiers in Ecology*, **6**: 346. We'd suggest that Strong is correct but too narrow in his interpretation. That is, ecologists do have values, but these are not necessarily those of environmentalists. Indeed, history reveals that ecology was as often motivated by exploitation as it was by preservation.
18. We don't expect our readers to set down this book at this critical juncture in order to contemplate their professional lives and then resign their academic positions to pursue volunteer work in the developing world. Although if such were to transpire, please let us know as it would be doubly heartwarming for us to have both stimulated the cultivation of a more meaningful life for a colleague and opened a desirable faculty slot for one of our talented students.
19. In spite of our accumulated mass of information, it seems that we never have the knowledge that we think we need. Perhaps a more explicit philosophy would foster better questions leading to a more effective development of knowledge. We might further suggest that some of the ideas we seek have long existed but scientists are disconnected from our own history. This is exacerbated by our new techniques of extracting literature through electronic means which filters out older material. In this regard, see: Evans, J. A. 2008. Electronic publication and the narrowing of science and scholarship. *Science*, **321**: 395–399.
20. Carlson A. 2000. *Aesthetics and the Environment: The Appreciation of Nature, Art, and Architecture*. New York: Routledge.
21. thinkexist.com/quotation/existentialism_means_that_no_one_else_can_take_a /198835.html, accessed October 20, 2008.
22. Crowell, S. 2004. Existentialism. *Stanford Encyclopedia of Philosophy*, plato.stanford.edu/entries/existentialism/, accessed October 20, 2008.
23. Lakoff and Johnson, *Metaphors We Live By*.
24. timeshighereducation.co.uk/story.asp?storyCode = 164451&sectioncode = 31 and nature.com/nature/journal/v421/n6925/full/421789a.html, accessed October 21, 2008.
25. theatlantic.com/issues/99aug/9908wdwtch.htm, accessed October 21, 2008.

12

The heroic handyman and the future
of ecology

Science is not at all immune to myth, by which we mean a traditional story that reveals an ideal toward which people aspire – as with the story of classical physics and its hero, Isaac Newton. The apple incident never happened, of course, but myths are not meant to be factual. The purpose of myth is not historical accuracy, but normative guidance. For many ecologists and philosophers, Newtonian physics has become the archetype of science.

We propose that ecology is far from the refined elegance of classical – might we say mythical – physics. Ecologists have no Platonic hero to distill the essence of reality into ideal, mathematical forms unsullied by the subjectivities of human perception and the complexities of chaotic contingencies. Instead, the mythic hero for ecology is an icon of pragmatism: the handyman.[1] This resourceful hero employs a flexible, analytical approach to fixing leaky pipes, rewiring a flickering light, or restoring a dilapidated porch. The handyman ecologist more typically resorts to analogical reasoning based on a period of apprenticeship from which highly contingent theories together with appropriate heuristics are combined to solve particular, local problems. No two houses are identical (note that ecology is etymologically rooted in the Greek, *oîkos*, meaning "house"). So it is incumbent upon the handyman to use experiments – like switch testing, valve manipulation, or frame fitting – to test hypotheses. The best-supported hypothesis leads to the operational tactic – a repair in the form of policy, intervention, or inaction.

Our metaphor can be extended to include the ecological rationalist, as well as the empiricist. Consider that plumbing problems demand different levels of knowledge and expertise. To install a sink requires only a rudimentary understanding of pipes and fittings, but fixing a leak in a high-pressure system on the space shuttle might well demand that one

194

know something about the chemistry of adhesives, the metallurgy of copper, the physics of fluids, the interdependent systems on board, the health risks of various interventions, etc. As such, the handy ecologist may draw upon rationalistically derived theories about function and then apply appropriate contingencies (Boxes 2.2 and 10.1). Or, to extend our metaphor a bit further, one might well compare the theoretician to a more elegant version of the handyman – the architect. Even this practitioner is seeking to solve a problem by drafting a blueprint in the form of a law, theory, or model.

So in either version of the handyman – the carpenter or the architect – if the product works to the satisfaction of the client (who may be a funding source, a government agency, a private industry, the public, or the scientific community), the practitioner moves on to the next problem. If the repair or blueprint fails, a new line of hypothesis development and testing must be pursued. The entire venture is unapologetically pragmatic, and whatever perspective works within the bounds set by the client's needs and desires is adopted. Of course, the perspective is constrained by aspects of the external world relevant to the slice of nature that one has chosen to engage. Using duct tape on high pressure steam pipes or drafting plans for a gravity-defying structure will assure that the world pushes back in a most decisive way. Moreover, it is worth noting that, to adapt a well-known saying: Architects without carpenters are lame, while carpenters without architects are blind.

Some handymen are better than others, and openness to alternative conceptualizations, alertness to new technical approaches, operational flexibility, truthfulness with self, and candor are often properties of the successful craftsman. So it is with ecologists when attempting to fix a broken habitat or a leaky theory. In this sense of determinable outcomes, ours is also a prescriptive philosophy in that the science of ecology will provide better answers if practitioners understand the necessity and value of a plurality of assumptions and methods.

The handyman as liberator

While analogizing science to the labor of a repairman or builder might seem to be mundane and even conservative, we would contend that the implications are potentially liberating and even radical. In effect, advocating pragmatism in the form of constrained perspectivism is a subversive philosophy for a no-longer subversive science.[2] Indeed, ecology has become, in many ways, a dreadfully safe and conventional science in which few revolutionary concepts are being proposed and there is little tolerance for ideas outside of the dominant paradigms (in the strictly Kuhnian sense).

We would argue that there has been little change in the framework of ecology in recent decades. The science is dominated by "normal" practices that fail to threaten the conceptual status quo. Tables of contents in contemporary ecology textbooks look remarkably similar to those of the 1970s (Box 12.1). And, one can argue that metapopulation theory, island biogeography, keystone species, or trophic cascades have not qualitatively changed anything from how ecologists saw nature 50 years ago.[3] Although there are surely intellectual clashes arising from the ambiguity of terms and conflicts among sub-disciplines, these conflicts generate more heat than light.[4] That is to say, there is little evidence that the borders of ecological science are being pushed, let alone breached. This lack of substantive change might be explained in various ways.

One could argue that ecology "got it right" a century ago, and no overthrow of fundamental concepts is necessary. However, such an interpretation would suggest that ecology is a remarkably special science in that physics, astronomy, chemistry, evolution, geology, embryology, physiology, and other disciplines have all seen conceptual revolutions.

Another possibility is that there have been paradigm shifts, but the nature of change has been quantitative rather than qualitative. That is, Kuhn's notion of scientific revolutions may be inapplicable to ecology.[5] Rather than change via the punctuated equilibrium of physics, ecology appears to have changed through prolonged gradualism. We are hard pressed to identify any major concept in ecology that has been utterly abandoned. Instead, ecologists seem to incorporate and modify earlier theoretical structures to serve contemporary needs. There have certainly been shifts in the relative emphasis placed on one or another perspective. But even these changes do not neatly accord with Kuhn's notion that change arises from within a discipline as "normal science" is no longer able to fill the gaping holes of a paradigm. Instead, the conceptual dynamics of ecology have often been driven from external, social forces. This phenomenon has been clearly documented by various historians, but the recent work pertaining to the role of religion in ecology is particularly compelling.[6] While incrementalism might explain some of ecology's history, the lack of a revolution might be indicative of a much more troubling possibility.

Box 12.1
So, what's new in ecology?

We systematically compared the contents and indices of: (1) Odum
E. P. 1971. *Fundamentals of Ecology*, 3rd edn. Philadelphia: Saunders
and (2) Townsend, C. R., M. Begon and J. L. Harper 2003. *Essentials of
Ecology*, 2nd edn. Malden, MA: Blackwell. Our major findings were:

1. Most of the entries that were unique to the 2003 text represented
   new terms for concepts that were found in the 1971 text (e.g.
   biodiversity).
2. The more recent text emphasized spatial aspects of ecology,
   but the difference was primarily a matter of framing older
   concepts in spatially explicit terms (e.g. metapopulations, island
   biogeography, and habitat fragmentation).
3. Townsend *et al.*'s text focused more on applied ecology and
   provided updated examples and methods, but the same
   environmental concerns were found in Odum's text (e.g. climate
   change, population viability analysis, habitat restoration,
   conservation biology, agricultural production, and fisheries
   management models).
4. The older text addressed essentially all of the major ecological
   concepts found in the modern version, although Townsend *et al.*'s
   text included refinements of these ideas (e.g. food web modeling
   of complexity and stability and the intermediate disturbance
   hypothesis).
5. Neither text made explicit mention of the "cutting edge"
   frameworks, including complexity theories (chaos, catastrophe
   theory, self-organized criticality), ecological topology, and the
   metabolic theory of ecology, although the empirical origins of
   MTE can be found in Odum's text.

The handyman saves ecology from
a premature demise

In 1996, John Horgan rattled the scientific world with his heretical
contention that we'd reached the end of science (at least as he and many
others understand this venture, which is rather more limited than how
pragmatism would define science – as we shall see).[7] So perhaps ecology,

a relative latecomer to science, was constructed on the foundations of past revolutions in physics, chemistry, and biology such that there are no major breakthroughs possible. Many of the major concepts of ecology are embedded in Darwin's *Origin of Species*, and a comparison of ecology textbooks from 1971 and 2003 revealed no radically new concepts in the last 30 years (Box 12.1). In a retrospective ten years after his original argument, Horgan seemed to be largely vindicated – between 1996 and 2006 there was little evidence to refute his claim that major discoveries had already been made, that the big scientific concepts were in place, or that science had become a matter of filling in the holes.[8]

Horgan makes a case for conceptual stasis, but his grim view reflects an understanding of science that is, at best, partial. Even if we grant that radical shifts in broad theoretical understandings are a thing of the past, this does not portend the end of science from the pragmatic perspective that we have developed as a philosophical framework for ecology.

Even Horgan admits that there are many new and exciting applications of science in fields where theoretical developments have ground to a virtual halt. He enthusiastically advocates a continuing investment in the sciences because of the potential for applications that could dramatically improve the quality of human life (Horgan also admits that his prediction of there being no scientific revolutions in the coming years could be wrong, and to withdraw support from the sciences would be tantamount to a self-fulfilling prophesy). In his estimation there remain a plentitude of deeply meaningful opportunities to relieve human suffering via the life sciences (e.g. treatments for AIDS, cancer, and mental illness), to improve the quality of life through the physical sciences (e.g. development of benign forms of energy production), to ameliorate environmental problems with earth system sciences (e.g. mitigation of global climate change), and to avoid killing one another through the social sciences (e.g. ending the use of war to solve human conflicts). But it seems that to Horgan, applied sciences are somehow disqualified as constituting real science.

We are suggesting, at least for ecology, that it would be a tragic error to equate "the end of science" with a period (even if indefinite in duration) during which upheavals in theory are lacking while revolutions in the uses of science are plentiful. So perhaps the revolution we need is not in how ecologists understand the world but in how they understand ecology (or science as a whole).[9] In this sense, pragmatism might represent something of a philosophical revolution which provides radically greater transparency, clarity, and integrity.

What would happen if we embraced pragmatism and admitted that human interests are at the root of both basic and applied research? We'll suggest one line of development that could arise from adopting such a view. If ecology (and perhaps all science) is grounded in human interests, including purely intellectual pleasures, then its practitioners might do well to adopt a practice from the realm of qualitative research.[10] That is, perhaps we ecologists should explicitly recognize the role of perspectives and divulge – in papers, proposals, textbooks, and syllabi – our ontological commitments, epistemological positions, and ethical motivations.[11] And we might well ask the same of funding sources – what are the philosophical perspectives from which a Request for Proposals is being issued by the National Science Foundation, the National Oceanic and Atmospheric Administration, the US Forest Service, the Bureau of Land Management, or a state wildlife agency? If science depends on the integrity of its practitioners and processes, then the interests that drive research and teaching in ecology should be made evident.

With this clarity would come transparency of our axiological considerations – the system of values that drives ecological investigations, including both basic and applied studies. While we would contend that funding should continue for lines of research driven by human curiosity, we should be transparent regarding our motivation. That is, we must recognize that we are intellectual hedonists and find the satisfaction of curiosity to be authentically pleasing. Moreover, we need to also admit that in the zero-sum game of public funding such forms of inquiry compete with investigations that are unambiguously aligned with imminent, collective human need for material well-being (e.g. clean air, fertile soil, productive fisheries, functional wetlands, etc.).

We expect that there would be many other important implications and applications of constrained perspectivism in the field of ecology. And whether or not ecologists should accept this philosophical framework requires the application of its own principles. That is, does constrained perspectivism address the desires and needs of our science – does adopting this perspective allow us to solve important problems of ecology itself? Is it *a* valuable tool (not *the* answer, for our appeal to pluralism is thoroughgoing) that can be applied to diminishing our confusion and destructive conflicts? We believe it is, and we invite our colleagues to consider the conditions under which this philosophical perspective is most useful.

Endnotes

1. We offer an apology for using this gender-exclusive term, and we recognize that there are plenty of female carpenters, plumbers, and architects. However, the use of "handyperson" was just so aesthetically awkward that we chose the more common, if less inclusive, "handyman."

2. Sears P. B. 1964. Ecology: A subversive subject. *BioScience*, **14**: 11–13; Shepard, P. and D. McKinley, eds. 1969. *Subversive Science: Essays Toward an Ecology of Man.* New York: Houghton Mifflin; Cuddington, K. and B. Beisner, eds. 2005. *Ecological Paradigms Lost.* Burlington, MA: Elsevier.

3. We concede that the incorporation of modern genetics into ecology has been a significant refinement, although not a conceptual breakthrough. In effect, the inclusion of genetic variation represents a de-idealization of ecological models and concepts, but the relationship between inherited traits and the environment is far from novel, being the basis of Darwinian evolution.

4. Stoll, M. 2006. Creating ecology: Protestants and the moral community of creation. In *Religion and the New Ecology: Environmental Responsibility in a World in Flux*, eds. D. M. Lodge and C. Hamlin. Indiana: University of Notre Dame, Indiana Press, 53–72; Cittadino, E. 2006. Ecology and American social thought. In *Religion and the New Ecology: Environmental Responsibility in a World in Flux*, eds. D. M. Lodge and C. Hamlin. Indiana: University of Notre Dame, Indiana Press, 73–115.

5. Kuhn, T. S. 1970. *The Structure of Scientific Revolutions.* Chicago: University of Chicago Press. We note, however, that what Kuhn took to constitute a "paradigm shift" might be rather different and more precise than what ecologists mean by this term. According to Beisner and Cuddington (2005. *Ecological Paradigms Lost, Routes of Theory Change*, 426), "The answer to the question of what ecologists mean when they use the phrase *paradigm shift* is likely to be: anything."

6. Stoll, Creating ecology; Cittadino, Ecology and American social thought.

7. Horgan, J. 1998. *The End of Science: Facing the Limits of Knowledge in the Twilight of the Scientific Age.* London: Abacus.

8. Horgan, J. 2006. The final frontier. *Discover*, October.

9. One might even make the argument that applied science is essentially the same as basic science. The difference is merely that the former operates under more highly constrained theory. That is, it could be the case that basic is equated with open-ended, broad theory and applied for special case theory. Under this view, the difference is a matter of limits under which the theories obtain.

10. Shank, G. D. 2002. *Qualitative Research: A Personal Skills Approach.* New York: Prentice-Hall.

11. Zellmer A. J., T. F. H. Allen and K. Kesseboehmer. 2006. The nature of ecological complexity: a protocol for building the narrative. *Ecological Complexity*, **3**: 171–182.

Glossary of philosophical terms

(These definitions were adapted from *A Dictionary of Philosophical Terms and Names*, philosophypages.com/dy/ix3.htm#p and *Dictionary of Philosophy*, itext.com/runes/f.html)

aesthetics   Branch of philosophy dealing with beauty or the beautiful. Its central issues include questions about the origin and status of aesthetic judgments: are they objective statements about genuine features of the world or purely subjective expressions of personal attitudes; should they include any reference to the intentions of artists or the reactions of patrons; and how are they related to judgments of moral value?

anti-realism   The view that we are not justified in accepting that the referents of scientific claims correspond with a mind-independent (objective) reality. As opposed to realism.

axiology   Branch of philosophy that studies judgments about value, including those of both aesthetics and ethics. The problems of axiology fall into four main groups: the nature of value, the types of value, the criteria of value, and the metaphysical status of value.

axiom   A proposition formally accepted without demonstration, proof, or evidence as one of the starting-points for the systematic derivation of an organized body of knowledge.

coherence theory   Belief that a proposition is true to the extent that it agrees with other true propositions. This view supposes that reliable beliefs constitute an inter-related system, each element of which entails every other. As opposed to correspondence theory.

consequentialism   A normative theory holding that human actions derive their moral worth from the outcomes or results that they produce. As opposed to deontology.

correspondence theory   Belief that a proposition is true when it conforms with some fact or state of affairs. As such, this theory holds that propositions are true when they correspond to reality. As opposed to coherence theory.

deduction   An analytical inference in which a conclusion follows necessarily from one or more given premises. As such, the conclusion will be of lesser generality than one of the premises. As opposed to induction.

deontology   A normative theory holding that moral worth is an intrinsic feature of human actions, determined by formal rules of conduct. Thus, deontologists like Kant suppose that moral obligation rests solely upon duty, without requiring any reference to the practical outcomes that dutiful actions may happen to have. As opposed to consequentialism.

dualism   Belief that the mental and the physical comprise fundamentally distinct kinds of entities. Primarily derived from Descartes and his followers; variations on this theme arise when dualists try to explain why events in the supposedly separate realms of mind and body seem so well-coordinated with each other.

empiricism   Belief that experience is the source of ideas and knowledge. More specifically, empiricism is the epistemological theory that genuine information about the world must be acquired by a posteriori means, so that nothing can be thought without first being sensed. As opposed to rationalism.

epistemology   Branch of philosophy that investigates the possibility, origins, nature, and extent of human knowledge.

ethics   Branch of philosophy concerned with the evaluation of human conduct with respect to judgments of approval and disapproval, rightness or wrongness, goodness or badness, virtue or vice, desirability or wisdom of intentions and actions.

existentialism   Belief that individual existence is primary, without there being any natural essence for human beings. Existentialists generally suppose that the fact of one's existence entails both unqualified freedom to make of oneself whatever one will and the responsibility of employing that freedom appropriately.

experientialism   Belief that neither subjectivism nor objectivism is a sufficient account of how we come to know. Experientialists hold that there is an external world, but truth is relative to understanding or perspective.

fallibilism   Belief that some or all claims to knowledge could be mistaken. Unlike a skeptic, the fallibilist may not demand suspension of belief in the absence of certainty. As opposed to foundationalism.

foundationalism   Belief that some or all claims to knowledge are not contextual or contingent, that there are absolute, usually a priori truths, about which there can be no mistake.

hypothesis   A proposition, tentatively put forward for the purposes of scientific explanation and subject to disconfirmation by empirical evidence.

induction   A form of reasoning from the observed to the unobserved such that the conclusion goes beyond what is formally contained in its premises. The truth of the premises merely makes it probable that the conclusion is true. As opposed to deduction.

knowledge   A justified true belief such that one knows a proposition if and only if: (1) one genuinely affirms the proposition, perhaps through acting upon it, (2) the proposition is true, and (3) one's affirmation is actually based upon its truth.

metaphysics   Branch of philosophy concerned with providing a comprehensive account of the most general features of reality. This field attends to questions about the nature and properties of that which exists.

mitigated realism   Belief that there are patterns in nature, existing independently of human perceptions, and these regularities are to some extent objectively knowable, but our perceptions are rooted in a cultural framework.

monism   Belief that there is but one fundamental reality, such that neither mind nor matter is ultimate but both properties may exist in particular things. As opposed to dualism.

naïve realism   An uncritical belief in an external world and the ability to know it, often ascribed to the "view of the man in the street."

naturalism   Belief that all objects, events, and values can be wholly explained in terms of factual and/or causal claims

	about the world, without reference to supernatural powers or authority.
nihilism	Rejection of belief in the existence of human knowledge, and the denial of the possibility of making any useful distinctions among things.
nominalism	Belief that only particular things exist. Nominalists hold that a general term or name is applied to individuals that resemble each other, without the need of any reference to an independently existing universal. As opposed to realism.
objectivism	Belief that real objects exist and are apprehended independently of the knowing mind, such that the truth of a claim is entirely unrelated to how people think or feel. As opposed to subjectivism; see also experientialism.
ontology	Branch of philosophy concerned with identifying, in the most general terms, the kinds of things that actually exist.
perspectivism	Belief that our knowledge is always situational within a physical, psychological, and cultural viewpoint, such that we can experience only a part of reality.
pluralism	Belief that because our knowledge is perspectival, claims of truth are invariably contextual and partial. As such, what is true depends on one's position in the world, but this means that a claim can be false within a domain or perhaps all domains.
postmodernism	Belief that confidence in the achievement of objective human knowledge based on reason (as associated with the Enlightenment) is mistaken. Postmodernists generally doubt the possibility of universal objective truth, reject dichotomies, and embrace the irony and particularity of language and life.
pragmatism	Belief that both meaning and truth can be explained in terms of the application of ideas or beliefs to the performance of actions that have observable practical outcomes. As such, the meaning of a proposition is its logical or physical consequences.
rationalism	Belief that reason is the only reliable source of human knowledge. The existence of a priori logical truths allows us to deduce inerrant knowledge.
realism	Belief that we are justified in accepting that the referents of scientific claims correspond to a mind-independent

	(objective) reality while recognizing fallibility (scientific methods may be mistaken) and approximation (most scientific knowledge is partially true).
realist	Belief that universals exist independently of the particulars that instantiate them. Realists hold that general terms signify actual features or qualities of the world. As opposed to nominalism.
reductionism	Belief that entities and explanations of one sort can be replaced systematically by entities and explanations of a simpler or more certain kind (e.g. the mental can be reduced to the physical or that the life sciences can be reduced to the physical sciences).
scholasticism	Belief that truth is knowable through a particular school of thought, particularly via rationalism.
skepticism	Belief that some or all human knowledge is impossible. Since even our best methods for learning about the world sometimes fall short of perfect certainty or completeness, skeptics argue that it is better to suspend belief than to rely on the dubious products of reason or experience.
solipsism	Belief that only oneself and one's own experiences are real, while anything else – a physical object or another person – is nothing more than an object of one's consciousness.
subjectivism	Belief that every object is created or constructed by the apprehender, such that truth depends upon the arbitrary expression of individual thoughts and feelings. As opposed to objectivism; see also experientialism.
theory	A systematic organization of knowledge with relatively high generality.
truth	See coherence theory, correspondence theory, and pragmatism.
utilitarianism	A normative theory holding that moral worth is a matter of producing the greatest good for the greatest number. Utilitarians typically identify happiness or pleasure (which may be understood as the higher pleasures, such as aesthetic enjoyment) as the favored consequence.

Index

Printed in the United States
By Bookmasters